T0331592

A Spotlight on the History of Ancient Egyptian Medicine

Global Science Education

Professor Ali Eftekhari

Series Editor

Learning about scientific education systems in the global context is of utmost importance for two reasons. Firstly, the academic community is now international. It is no longer limited to top universities, as the mobility of staff and students is very common even in remote places. Secondly, education systems need to continually evolve to cope with market demand. Contrary to the past, when the pioneering countries were the most innovative, now emerging economies are more eager to push the boundaries of innovative education. Here, an overall picture of the whole field is provided. Moreover, the entire collection is indeed an encyclopedia of science education and can be used as a resource for global education.

A Spotlight on the History of Ancient Egyptian Medicine

Ibrahim M. Eltorai

CRC Press
Taylor & Francis Group
Boca Raton London New York

CRC Press is an imprint of the
Taylor & Francis Group, an **informa** business

CRC Press
Taylor & Francis Group
6000 Broken Sound Parkway NW, Suite 300
Boca Raton, FL 33487-2742

Visit the Taylor & Francis Web site at
http://www.taylorandfrancis.com

and the CRC Press Web site at
http://www.crcpress.com

Contents

Introduction

THIS BOOK PROVIDES THE reader with an outline of ancient Egyptian civilization, history, and culture. It then reviews the ancient Egyptian understanding of human health and disease and medical and herbal treatments for various conditions based on primary sources found in ancient papyri. The reader will also gain insight into the influence of ancient Egyptian medical knowledge on later civilizations, including ancient Greek and Islamic scholars of the Middle Ages. Two chapters focus on the ancient Egyptian understanding and treatments of diseases of the heart and blood vessels.

This educational text would be of interest to medical and nursing students as well as students interested in the history of science and human civilizations.

The first edition of this book was self-published in Egypt in the 1960s, and the publisher is no longer in business. The title of the first edition was: *A Short Sojourn in Pharaonic Egypt, A Precis of Its History, and an Exposé of Ancient Egyptian Medicine. With Special Emphasis on Cardiovascular Knowledge* by Ibrahim M. Eltorai, MD, Ch.M., FACS.

introduction

Acknowledgments

For the first edition:
My deepest gratitude goes to the librarians of:

- Cairo Museum Library

- Institut Français d'Archéologie Orientale

- The Vatican Library

- Le Bibliothèque Nationale, Paris

- The Wellcome Institute of History of Medicine, where most of the time was spent for research

My heartfelt thanks go to:

- Sir Zachary Cope,[†] past Honorary President of the British Society of History of Medicine, for reporting on the manuscript in spite of his imperfect health

- Doctor Khalid El-Dissouky, Professor of Archeology at Ain-Shams University, for revising the manuscript and his most valuable advice

- Professor Paul Ghalioungui[†] for his valuable suggestions and advice

---

[†] Deceased.

- Mr. MH Abdul Rahman, the Director of the Cairo Museum, for verification of the history portion and the chronology of the dynasties

- Professor Vivi Täckholm,[†] of Cairo Faculty of Science, for her interpretations of ancient Egyptian flora

- Mr. Kamal Zaki for the production of offset prints

For the second edition, I am most grateful to:

Sir Magdi Yacoub, Fellow of the Royal College of Surgeons, Licentiate of the Royal College of Physicians, and Fellow of the Royal Society of Medicine. He was so gracious and kind to review this text in spite of his extremely busy schedule. I am honored to have Sir Yacoub's advice and suggestions.

Dr. Fayza Haikal, Professor of Egyptology at the American University in Cairo, distinguished researcher and scholar of Egyptology, for reviewing the history section of the manuscript and for the advice.

Dr. Wafaa Elseddeek, Director of The Egyptian Museum, for permission to publish the pictures from the museum.

Elhami Boulos, director of Al-Hadara Publishing, Cairo.

My gratitude to my wife, Salwa, for her encouragement and patience over the years while preparing this manuscript.

I thank Ms. Leslie K. Shimoda for her hard work and dedication to the preparation of the manuscript.

I also thank my grandson, Adam Eltorai, M.D., Ph.D., for all his efforts in getting the book published, and my son, Mahmoud Eltorai, M.D., for his assistance with the final edits to the manuscript.

Ibrahim M. Eltorai, M.D., Ch.M., F.A.C.S.

Author

Ibrahim M. Eltorai completed his medical education at Cairo University in Egypt and then completed postgraduate training in neurosurgery and cardiovascular surgery at multiple European institutions. He joined the faculty of the school of medicine at Cairo University and became Professor of Surgery. He has been published nearly 150 times in Egyptian and international medical journals. When he moved to the United States, he joined the Spinal Cord Injury Service at the Long Beach V.A. Medical Center, where he worked for more than 30 years until his retirement in 2004.

Dr. Eltorai published numerous clinical articles in the field of spinal cord injury (SCI), edited the book *Emergencies in Chronic Spinal Cord Injury Patients*, and is the author of the book *Rare Diseases and Syndromes of the Spinal Cord*. He was a founding member and President of the American Paraplegia Society, as well as a Fellow of the International Spinal Cord Society. Dr. Eltorai's research on the evolution of SCI medicine resulted in a comprehensive library, which is currently located at the Palo Alto V.A. Medical Center.

CHAPTER 1

Why Study the History of Medicine?

L ET US READ WHAT Fuller (1902) said about studying medical history:

> The history maketh a young man to be old without either wrinkles or grey hair; privileging him with the experience of the age without either the infirmity or inconveniences thereof. Yeah, it not only maketh things past, present; but enableth one to make a rational conjecture of things to come. For this world affordeth no new accidents, but in the same sense wherein we call it a new moon, which is the old one in another shape; and yet no other than that hath been formerly. Old actions return again, furbished over with some new and different circumstances.

Some may not agree and may consider these ideas old-fashioned, so I quote from Goethe:

> It is a great pleasure to return to the spirit of the former days, and to see what a wise man has thought before us.

> JOHANN WOLFGANG VON GOETHE, *FAUST*, PART 1 (TRANSLATED)

Any physician in the world has but one aim, which is to cure or improve the patient's condition and to achieve for him a healthy life. One has to use a remedy, as effective as possible, no matter where it originated, to benefit all of humanity.

One of the foremost pioneers of Anglo-American medicine, Sir William Osler, wrote the following in the preface to *History of Medicine* by Dr Neuburger (1910) of Vienna:

> The subject 'History of Medicine' has three relations: For the student the educational aspect is of incalculable value, since medicine is best taught from the evolutionary standpoint. What a help it is to give early in his career a clear view of the steps by which our present knowledge has been reached!
> Secondly as a study, the history of that branch of science, which had to do with healing, has peculiar attractions. With foundations in anthropology medicine has close affiliations with most of the theologies, many of the philosophies, and with the 'pseudosciences' of alchemy and astrology. To trace its evolution, to study the relations which it has borne to the intellectual movements at different periods, is the work of scholars trained in modern methods of research ... Even more striking has been the growth of the subject in its third relation—a useful pastime for the leisure moments of busy men who take an interest in the history of the profession, local or general.

And from Robert Browning (1983), we quote:

> 'Tis time
> New hopes should animate the world, new light
> should dawn from new revealings to a race,
> weighed down so long, forgotten so long.

Since the fundamental problems of disease are almost the same across the ages, early ideas in medicine should be revisited. In fact, many of these ideas have been elaborated in modern science. It is experimentation and investigation that will prove the ancient medical hypotheses or disprove them. It is not infrequent to find that modern science has transferred the crude knowledge of the ancients into scientific principles. For example, antibiotics were only discovered in the beginning of the twentieth century. However, if scientists had learned ancient history and investigated it scientifically and without prejudice, antibiotics may have been discovered much earlier. Over 4,000 years ago the ancient Egyptians used empirically moldy wheaten loaf, onions, garlic, radish, spoilt meat, yeast, rotted stems of lotus flowers, and different types of soil. Examples will be presented in chapter "Ancient Egyptian Pharmacology". Other ancient medical practices, such as those of the Chinese and the Indian people, are now recognized in the Western World (Twentyman, 1954; Stever and Saunders, 1959). We are fortunate to have modern scientific disciplines such as molecular biology, genetics, microbiology, which can be used to evaluate ancient medical remedies.

Now we know that medicine is as ancient as human history; they go hand in hand with human civilization (Stauffer, 1943). Although very primitive man struggled through countless dark labyrinths toward enlightenment, Egyptian history gives us the most remote records on the antiquity of medicine. Other ancient nations such as Babylonia, Assyria (Budge, 2000), Persia, India, and China also contributed to laying down the foundation of medicine. Later, the Greeks, Romans, Hebrews, and Arabs were handed the lit torch and they made it brighter and made the way clearer. Just like what we have in modern times, our ancestors had in ancient times the same aim, namely the protection of the human race.

It is clear now that the study of the history of medicine is quite a useful one, although unfortunately it was relatively neglected until the twentieth century. Fortunately, there are eminent scholars who combined Egyptology with the history of medicine in ancient Egypt. Just to mention a few: Paul Ghalioungui, Hassan Kamal, Hanifa Moursi, Naguib Riad and Fayyad- all from Egypt, J. Worth Estes from England, Bruno Halioua and Bernard Ziskind from France, Cornelius Stetter from Germany. In this text, the focus will be on ancient Egyptian medicine, particularly highlighting the history of cardiovascular diseases, which remain very common in modern times as outlined in the chapter on their epidemiology. As previously mentioned, the history of medicine and the history of civilization are intertwined. It is therefore of great importance to students of medicine to understand Egyptian history. The study of ancient Egyptian history has attracted many world scientists, who coined the term Egyptology. This is a vast branch of history, which demands in-depth specialization. However, for our purpose, the history of ancient Egyptian medicine will be summarized briefly in the following pages.

HISTORICAL OVERVIEW OF ANCIENT EGYPT

The history of medicine in ancient Egypt cannot be complete without a short description of the land of the pharaohs, in which many mysterious and majestic developments have taken place over 5,000 years. For thousands of years, the majestic pyramids have been a reminder of the golden age of ancient Egyptian civilization. The writings from the Bible, Homer, Herodotus, the Hellenic philosophers, and historians described the legacy of ancient Egypt long after the hieroglyphics had become undecipherable. While other countries were still in the dark ages, the Egyptians were making advancements in science, engineering, art, metaphysics, and mysticism. Subsequent cultures expressed wonder about these achievements but did not have full understanding of this civilization because its recorded language remained mysterious. Significant progress was made in understanding Egyptian civilization when the trilingual Rosetta Stone (see Figure 1.1) in 1822 provided the key to the forgotten language of the pharaohs. This critical step was credited to the French Egyptologist Jean-François Champollion (1790–1832) (1824). Egyptologists could then bring to light the treasures hidden for thousands of years in the pyramids, tombs, papyri, temples, inscriptions, etc. It is due to the desert and the dry and almost rainless climate of Egypt that its past civilization was preserved to a great extent. The imposing architecture; the decorative ingenuity and fidelity to nature in artistic representation; the advanced chemical technology; the early understanding of mathematics and geometry applied to the design of structures of every kind; the copious literature on

FIGURE 1.1   Rosetta Stone (commons.wikimedia.org).

the philosophy of religion, astronomy, poetry, science, and medicine have exceeded the scientists' expectations, especially when their dates of origin in antiquity are taken into consideration. All of these discoveries have attracted countless scientists, especially from Europe and America, who work in diverse fields. This is evidenced by the immense literature, monuments, and artifacts available in Egypt, as well as those housed in many museums and historical libraries around the world. (See the list in *Ancient Egypt* by Oakes and Gahlin.)

In his book *History of Egypt*, Professor James Henry Breasted (1948) wrote:

> As the Nile poured its life-giving waters into the broad
> basin of the Mediterranean, so from the civilization of the
> wonderful people who so early emerged from barbarism
> on the Nile shores, there emanated and found their way
> to Southern Europe rich and diversified influences of culture to which we of the Western World are still indebted.

It is impossible to follow the history of science without knowing the chronology of those old times. A brief account is hereby given.

OUTLINE OF ANCIENT EGYPTIAN HISTORY

The story of Egypt begins when the first settlers descended into the Nile Valley before 7000 B.C. Thus started that vast era, termed Prehistory, during which the Egyptians began their long journey toward civilization. This early stage of Egyptian history with its various cultures is beyond the scope of this book. Instead, this outline will summarize the history of ancient Egypt, starting with the early dynastic period.

Manetho, an Egyptian priest who lived in the third century B.C., wrote a history from priestly records in which he divided the kings of Egypt into 30 dynasties, i.e. royal families. This method proved so convenient that it is still retained by present historians. The latter, however, divide the dynasties into groups corresponding with periods. These periods can be summarized as follows:

I. Archaic Period (Dynasties I–II, c. 3100–2686 B.C.): This history of Egypt started when a king from the South, named Menes (probably Narmer), united the whole country under his scepter. This occurred around 3100 B.C. He and his successors created a strong, centralized administration at the head of which they ruled. In addition, Menes is traditionally known to have founded a new capital in a place where Upper and Lower Egypt met, which he called Ineb-Hedj or "The White Wall". Later it became known as Memphis. During this early period Egyptian culture matured rapidly and the foundations of the Egyptian civilization were firmly established.

II. Old Kingdom (Dynasties III–VI, c. 2686–2181 B.C.): The kings of this epoch made Memphis their capital. In this period, which is sometimes termed the Pyramid Age, Egypt reached the peak of prosperity, splendor, and cultural achievement. Zoser, the greatest king of Dynasty III, built the step pyramid and its complex at Saqqara – a landmark in the history of stone architecture (see Figure 1.2). Its architect was probably Imhotep, the famous vizier, sage, and physician.

In Dynasty IV, the divinity, and hence the authority, of the king became absolute as never before or since. A series of pyramids were built that culminated in the Great Pyramid of Khufu (Gk.: Cheops) at Giza (c. 2650 B.C.). Other manifestations of civilization were achieved, e.g. sculpture, painted reliefs, furniture, and jewelry (see *The Great Pyramid* by P. Smyth, 1995). In his book *The Secrets of the Great Pyramid*, P. Tombkins (1996) gives a description of the Great Pyramid, its structure, engineering, scientific theories, astronomical observatory, and astrological observatory.

FIGURE 1.2   Saqqara/Sakkara Step Pyramid (commons.wikimedia.org).

He gives details about the great technological aspects of the pyramids, especially the geometrical, geological, and geographical characteristics, which were amazing 4,000 years ago and still offer many mysteries.

Dynasty V witnessed the rise of the cult of Re, the sun-god (and its priests), to dominance. Temples for its worship were built as well as the pyramids and memorial temples of the pharaohs. For various reasons, the power of the king, and the keystone in the Egyptian system, began to suffer gradual decline.

The kings of Dynasty VI sent expeditions to trade with Punt* (on the African coast near the straits of Bab el Mandeb) and explore Nubia. The walls of the inner rooms of their pyramids were covered with religious inscriptions known as the Pyramid Texts. The decline of the king's powers continued on the one hand, while the local princes became increasingly independent on the other. This situation reached its climax late in the long reign

---

* Punt: far away to the southeast of Egypt in the latitude of Eritrea and Somaliland was a mysterious land known to the Egyptians since the fifth dynasty. It produced ebony, resin, and other goods. It was visited by Egyptian trading expeditions. The Queen of Punt was depicted in a relief from Queen Hatshepsut;s temple at Deir el-Bahri.

(94 years) of Pepi II. By the time of his death, the glorious epoch of the Old Kingdom came to an end.

III. First Intermediate Period (Dynasties VII–X, c. 2181–2040 B.C.): The first tragic collapse of the monarchial system of the Old Kingdom was followed by a period of political confusion and social upheaval. In Memphis presided the shadowy kings of Dynasties VII–VIII, but they had little effective control over Egypt. The country was divided into conflicting little princedoms, which fought with one another, and Egypt was plunged into a kind of civil war. Taking advantage of this chaotic situation, foreign forces infiltrated from the south and the northeast. The state of the country deeply shocked the conscience of Egyptian thinkers, who had believed that their order was divine, hence eternal. This is clearly reflected in the remarkable literature that flourished in this period.

The kings of Dynasties IX and X, who presided in Nen-nesut (near Fayum), endeavored to have full control of Egypt, but they achieved only limited success. They collided with the powerful princes of Thebes, who had declared their independence (Dynasty XI). Eventually, one of the Theban princes, namely Mentuhotep II, vanquished the northern kings, reunited Egypt, and founded the Middle Kingdom.

IV. Middle Kingdom (Dynasties XI–XII, c. 2040–1768 B.C.): The kings of Dynasty XI made Thebes their mother city and the capital of their kingdom. Mentuhotep II, the most outstanding figure of the Dynasty, conducted a series of successful campaigns in which he drove out the foreign intruders, whether Asiatics or Nubians. He built the most remarkable funerary complex on the western side of Thebes.

When Dynasty XI ended in confusion, the throne went to Amenemhat I, the founder of Dynasty XII, c. 1991 B.C. He chose a site near Fayum where he founded a new capital called Ithettawy, the Controller of the Two Lands. The kings of this Dynasty, named Amenemhats and Senuserts, rebuilt a strong central administration, settled land disputes, and promoted agricultural prosperity. Some of the kings, particularly Senusert III, subdued Nubia as far as the Second Cataract and built a series of forts and trading stations. Senusert III also raided Palestine in order to secure the northeastern frontiers. Internally, he abolished the authority of the local rulers and put an end to their power.

The Middle Kingdom witnessed the flourishing of art and architecture once more. In many respects, this was the golden age of classical ancient Egyptian literature, especially short stories.

V. Second Intermediate Period (Dynasties XIII–XVII, c. 1768–1567 B.C.): During the times of Dynasties XIII and XIV, the power of the kings deteriorated and the machinery of the state began to run down. Egypt fell into a state of confusion and disunity comparable with the First Intermediate Period. Consequently, hordes of Asiatics, mainly Semites, infiltrated into Egypt and gradually established themselves in the Eastern Delta. In due course, their chieftains gained prominence in Lower Egypt and founded a capital in the Eastern Delta, namely Avaris.* The Egyptians termed these chieftains Hekraw-khasut, i.e. rulers from the foreign lands, from which the name Hyksos was derived.

The Hyksos' rule, which began in about 1720 B.C., constituted Dynasties XV and XVI. At one stage, they seem to have controlled the whole of Egypt, although that was only for a relatively short time. It is evident from the Egyptian literature from this epoch that the national feelings were wholly opposed to the foreign overlords.

---

* Bietak, Manfred: Avaris, capital of the Hyskos. London: British Museum, 1996.

Again in Thebes rose a family that started the war of liberation against the foreign rulers. Seqenenre was the first prince to fight them; he was killed in action. Kamose, his eldest son, took over and continued the war and drove the Hyksos as far as Avaris itself. The final victory, however, was destined for his brother Abmose. He captured Avaris, threw the enemy out of Egypt, and even pursued them into Palestine in order to defeat them completely. With the reign of Abmose began the third great epoch of Egyptian history.

VI. New Kingdom (Dynasties XVIII–XX, c. 1570–1070 B.C.): The New Kingdom witnessed the pinnacle of Egypt's political power and imperial expansion. Realizing that the conventional frontiers of Egypt were not secure, the pharaohs pushed their boundaries deep into Western Asia as far as the Euphrates and in Nubia as far as the Fourth Cataract. Tuthmosis I first carried out the policy of imperial expansion. His grandson Tuthmosis III conducted no less than 16 campaigns in Palestine and Syria in order to consolidate the Egyptian rule there. This policy was consistently followed by his successors Amenophia II and Tuthmosis IV. As a result, the wealth of Western Asia and Nubia poured into Thebes, the Imperial Capital. A series of magnificent buildings, temples, palaces, etc. rose on both sides of Thebes. The kings hewed their beautifully decorated tombs in the cliffs of western Thebes in the Valley of the Kings. The nobles and high officials of the state cut their tombs in the same cliffs not far from those of their royal masters. The chapels were wonderfully decorated with scenes recording the different activities of daily life and scenes of religious and funerary beliefs.

Amenophis III (the Magnificent) enjoyed all the splendor and luxury of the empire but unfortunately did very little to maintain it. When his son Amenophis IV, better known as Akhenaton, ascended to the throne, he led a religious revolution against the imperial god Amun and its hierarchy (see Akhenaton in the chapter on religion). Then he broke completely with all gods and proclaimed Aton (manifested in the sun-disc) as the sole god. He left Thebes and moved to a newly built city near Menya. His revolution was also against the long-established conventions of art and even language.

Akhenaton became so preoccupied with his revolution that Egypt's hold on Syria and Palestine slackened and the local princes conflicted with one another. The Hittites, in Asia Minor, took over northernmost Syria and imposed their rule on Amurru, in central Syria; vast territories of the empire were thus lost. During the religion revolution, which is well documented by Breasted, poems (psalms) pertaining to the new religion were written. A few of them are given here (Breasted, 1948):

THE WHOLE CREATION

> How manifold are all thy works!
> They are hidden from before us,
> O lord, how manifold are thy works!
> While thou wast alone:
> O thou sole god, whose powers no other possesseth.*
> Thou didst create the earth according to thy desire.
> Men, all cattle large and small,
> All that are upon the earth,
> That fly with their wings.
> The countries of Syria and Nubia,
> The land of Egypt;
> Thou settest every man in his place,
> Thou suppliest their necessities.
> Every one has his possessions,
> And his days are reckoned.

---

* The other hymns frequently say, "O thou sole god, beside whom there is no other."

Their tongues are diverse in speech,
Their forms likewise and their skins,
For thou divider, hast divided the peoples.
In wisdom hast thou made them all;
The earth is full of thy creatures.

(PSALM 104:24)

CREATION OF MAN

Thou art he who createst the man-child in woman,
Who makest seed in man,
Who giveth life to the son in the body of his mother,
Who soothest him that he may not weep,
A nurse [even] in the womb.
Who giveth breath to animate every one that he
  maketh.
When he cometh forth from the body,
… on the day of his birth,
Thou openest his mouth in speech,
Thou suppliest his necessities.

CREATION OF ANIMALS

When the chicklet crieth in the egg-shell,
Thou givest him breath therein, to preserve him alive.
When thou hast perfected him
That he may pierce the egg,
He cometh forth from the egg,
To chirp with all his might;
He runneth about upon his two feet
When he hath come forth therefrom.

Internally, discontent was increasing and economic decline was
setting in, so Akhenaton had to compromise with his opponents

(Abd-Ur-Rahman, 1959, Weigall, 1922). When Tutankhamun assumed the throne, he repudiated the cult of Aton and returned to the old beliefs. However, the affairs of Egypt were set right again by General Horemhab, who became pharaoh at the end of the Dynasty. Succeeding him to the throne was a military colleague of his, Ramses I, who founded Dynasty XIX.

The burden of recovering parts of the lost territories fell heavily on the shoulders of Sethos I and his son Ramses II. Individually, they fought the Hittites and achieved a certain amount of success. Ramses met them in the notorious battle of Qadesh in which his army almost faced defeat, but thanks to the personal valor of the King, Egypt was saved. In the 21st year of his reign, he signed a peace treaty with the Hittites. To foster this treaty, he married the daughter of the Hittite king in the 34th year and another daughter later. His son Merenptah, when king, faced an invasion of the Sea Peoples from Libya, which he successfully beat off.

During the reign of Ramses III, the second pharaoh of Dynasty XX, the threat from the Sea Peoples became imminent. In the eighth year of his reign, he fought a remarkable battle against them and destroyed their fleet in a naval battle. In the eleventh year, he crushed the Libyans, who threatened the western frontiers and with whom he had fought before in the fifth year.

Ramses III was probably the last of the great pharaohs. He was succeeded by eight weak pharaohs, each one calling himself Ramses. The state of the country went from bad to worse. The high priests of Amun at Thebes grew increasingly powerful until one of them, namely Herihor, assumed the royal titles and ruled Upper Egypt. Meanwhile, a local prince, named Nesubanebded, ruled Lower Egypt from Tanis in the Eastern Delta – thus ended the era of the New Kingdom.

VII. Late Period (Dynasties XXI–XXXI, c. 1085–332 B.C.): This was a period of general decline, halted at intervals by occasional kings of outstanding character. Nesubanebded and Herihor seemed to have reached a mutual agreement through which the former was recognized as the sole pharaoh of Egypt, while the latter and his successor became the high priests of Amun and the effective rulers of Upper Egypt.

When the last Tanite king died, the throne passed into the hands of a Libyan tribal chief, namely Sheshonq I, who established the Libyan Dynasties XXII–XXIII. He invaded Palestine and subdued the kingdoms of Israel and Judah.

Dynasty XXIV comprises only two kings who ruled parts of Egypt from Sais in the Western Delta. This short-lived Dynasty was ended by a Nubian invasion. Piankhy, a Napatan king, led an army into Egypt, entered Thebes, then captured Memphis, and so founded Dynasty XXV.

In the final years of Dynasty XXV, the Assyrian king Ashurpanibal invaded Egypt. He drove out the Nubian family and sacked Thebes.

An able prince from Sais, namely Psammetichus, managed to unite Egypt under his scepter. He then succeeded in driving the Assyrians out and established Dynasty XXVI, commonly known as the Saitic Period. During the reign of the Dynasty, Egypt enjoyed a spell of prosperity and internal security. But this period came to a tragic end when the Persian army led by Cambyses invaded Egypt in 525 B.C. and Dynasty XXVII, of Persian rulers, was founded.

The Egyptians revolted against the Persians and recovered their independence for a short time during 400–341 B.C. (Dynasties XXVIII–XXX). However, the Persians regained control of Egypt, but this did not last for more than nine years (Dynasty XXXI). In the year 332 B.C., Alexander the Great marched into Egypt and put an end to the Persian occupation. This began an entirely new era in Egyptian history. The dynasties are listed in Table 1.1 (Casson and Kriegger, 1965; Posener, 1959).

The Greco-Roman periods will not be presented in this short historical review, since it focuses on pharaonic Egypt. For more details about dynasties, see Oakes and Gahlin, Nunn, and other texts on ancient Egypt.

TABLE 1.1   A Chronology of Ancient Egyptian Dynasties

	Time Period Approximate Dates (B.C.)		
Prehistoric Period			
Paleolithic Age	Before 7000	Lower	Clactonian-Abbevillian
		Middle	Acheulean
		Upper	Mousterian and Levalloisian
Neolithic Period	7000–5000	South	Tasian
		North	Merimdian, Omarian
Cahalcolithic or Predynastic Age	5000–3100	South	Badarian
			Nagada I: Amratian
			Nagada II: Gerzean
		North	Meadi and Heliopolis

The end of the prehistoric period (sometimes called Pre-Thinite or Protohistoric Period) and the Thinite Period form the Archaic Period (v. Origins).

	Approx. Dates(B.C.)	Dynasty	Rulers
Early Dynastic Period	3100–2890	1st	Menes, Djer
	2890–2686	2nd	
Old Kingdom	2686–2181		
	2686–2625	3rd	Zoser
	2625–2500	4th	Snefru, Cheops, Chephren, Mycerinus
	2500–2350	5th	
	2350–2181	6th	Memphite
First Intermediate Period (revolution)	2181–2040		
	2181–2173	7th–8th	
	2160–2040	9th–10th Dynasties	Herakleopolitan in the North
	2133–1991	11th	Theban (the Antefs) beginning in the South, Mentuhotep I

(Continued)

TABLE 1.1 (CONTINUED)   A Chronology of Ancient Egyptian Dynasties

	Approx. Dates(B.C.)	Dynasty	Rulers
Middle Kingdom	2040–1786		
	1938–1759	12th	Ammenenemes and Sesostris
Second Intermediate Period	1786–1567		
	1786–1633	13th	Sebekhotep and Neferhotep
	1675–1630?	14th	
	1674–1567	15th–16th	Hyksos in the North
	1650–1567	17th	Theban in the South: Kamose
New Kingdom (second Theban empire)	1570–1070		
	1570–1293	18th	Amosis, Amenophis, Thutmosis, Hatshepsut, Akhnaton, Tutankhamun, Horemheb
	1293–1185	19th	Seti I, Ramses I-II, Merneptah (Ramesside Dynasties)
	1185–1070	20th	Ramses III–XI (Ramesside Dynasties)
Third Intermediate Period	1070–664		
	1070–945	21st	Tanis in the North: Psusennes
			Priest Kings in the South
Libyan Period	945–730	22nd	Sheshonq at Bubastis
	817–730	23rd Dynasty	Petubastis at Tanis
	720–714	24th Dynasty	Bocchoris at Sais
Ethiopian or Kushite Period	747–656	25th	Taharqa

(Continued)

TABLE 1.1 (CONTINUED) A Chronology of Ancient Egyptian Dynasties

	Approx. Dates(B.C.)	Dynasty	Rulers
Late Period	664–332		
Saite Period	664–525	26th	Psammetichus, Necho, Apries, Amasis
1st Persian Domination	525–404	27th	Cambyses, Darius, Xerxes
Last Native Kings	404–399	28th	at Sais
	399–380	29th	at Mendes
	380–343	30th	at Sebennytos (Nectanebo I–II)
2nd Persian Domination	343–332	31st	
Alexander the Great	332–323		
Hellenistic Period	332–305		
Ptolemaic or Lagid Period	305–30		Ptolemy, Cleopatra
Roman Period, Coptic or Byzantine Period	30B.C.– 641A.D.		
Arab Conquest	641 A.D.		

Ancient Egyptian excavation artifacts are found in museums all over the world, in addition to private collections. Oakes and Gahlin tabulated details of world museums having Egyptian archaeological materials of different categories. There are Egyptian archaeological valuables in 27 countries and in 123 cities. The most numerous are in collections in France, Germany, Italy, Switzerland, the United Kingdom, and the United States of America.

Some of the best known places are the following:

In France:

Paris – the Louvre, the Bibliothèque Nationale

Strasbourg – Institut d'Egyptologie

In Germany:

    Berlin – Berlin Museum

    Leipzig – Ägyptisches Museum

    Heidelberg – Ägyptologisches Institut der Universität

    Munich – Staatliche Sammlung: Ägyptischer Kunst

    Tübingen – Ägyptologisches Institut Universität

In Italy:

    Rome – Museo Barracco, Musei Capitolini

    Turin – Museo Egizio

In Switzerland:

    Basel – Museum für Völkerkunde

    Geneva – Musée d'Art et d'Historie

In the United Kingdom:

    London – British Museum, Petrie Museum of Egyptian Archaeology

In the United States:

    Art museums in 30 big cities, notably the Metropolitan Art Museum in New York and the museum at the University of Chicago.

The Cairo Museum in Egypt has the largest collection.

SUGGESTED READINGS ON ANCIENT EGYPTIAN HISTORY

Aldred, C. *The Egyptians*, 3rd ed., revised and updated by Aidan Dodson. London: Thames and Hudson, 1998.

Bagnall, R.S. *Egypt in Late Antiquity.* Princeton, NJ: Princeton University Press, 1993.

Baines, J. and Malek, J. *Atlas of Ancient Egypt*. New York: Checkmark Books, 1980.

Bard, K. (ed.). *Encyclopedia of the Archaeology of Ancient Egypt*. London and New York: Routledge, 1999.

Berman, L.M. and Lettelier, B. *Pharaohs of Egypt: Treasures of Egyptian Art from the Louvre*. Oxford University Press, 1996.

Bowman, A.K. *Egypt after the Pharaohs*. University of California Press, 1986.

Brewer, D.J. and Teeter, E. *Egypt and the Egyptians*. Cambridge University Press, 1999.

Clayton, P.A. *Chronicle of the Pharaohs: The Reign-by-Reign Record of the Rulers and Dynasties of Ancient Egypt*. London: Thames & Hudson, 2006.

Dawson, W.R. and Uphill, E.P. *Who Was Who in Egyptology*, 3rd ed., revised by M. L. Bierbrier. London: Egypt Exploration Society, 1995.

Emery, W.B. *Archaic Egypt*. Harmondsworth: Penguin Books, 1961.

Grimal, N. *A History of Ancient Egypt*. Oxford: Blackwell Publishers, 1992 (Paperback edition, 1994).

Hancock, G. and Bauval, R. *The Message of the Sphinx*. New York: Three Rivers Press, 1996.

Harris, J.R. (ed.): *The Legacy of Egypt*. Oxford Clarendon Press, 1971.

Hawass, Z.A. *Tutankhamun: The Golden King and the Great Pharaohs*. Washington DC: National Geographic Society, 2008.

Hayes, W.C. *The Scepter of Egypt: A Background for the Study of Egyptian Antiquities in the Metropolitan Museum of Art*, 2 Vols. New York: Metropolitan Museum of Art (Volume 1, *From the Earliest Times to the End of the Middle Kingdom*. 1953; Volume 2, *The Hyksos Period and the New Kingdom*. 1959).

Helck, W., Otto, E., and Westendorf, W. (eds.). *Lexikon der Ägyptologie*, 7 Vols. Wiesbaden: Harrassowitz, 1972–1992.

Hoffman, M. *Egypt Before the Pharaohs: The Historical Foundations of Egyptian Civilization*, Rev. ed. Austin, TX: University of Texas Press, 1991.

Janot, F. and Hawass, Z. *Royal Mummies: Immortality in Ancient Egypt*. New York: White Star Publishers, 2010.

Kemp, B. *Ancient Egypt: Anatomy of a Civilization*. London: Routledge, 1989.

Kitchen, K.A. *Pharaoh Triumphant: The Life and Times of Ramesses II*. Warminster: Aris & Phillips Ltd., 1982.

Kitchen, K.A. *The Egyptian Nineteenth Dynasty*. Warminster: Aris & Phillips, 1977.

Kitchen, K.A. *The Third Intermediate Period in Egypt (1100–650 B.C.)*, 3rd ed. Warminster: Aris & Phillips Ltd., 1995.

Manley, B. *The Penguin Historical Atlas of Ancient Egypt.* London: Penguin Books, 1996.

Meyers, E.M. (ed.). *Oxford Encyclopedia of Archaeology in the Ancient Near East*, 5 Vols. Oxford University Press, 1997.

Osman, A. *Moses and Akhenaten: The Secret History of Egypt at the Time of the Exodus.* Boston: Bear & Company, 2002.

Quirke, S. and Spencer, J. *The British Museum Book of Ancient Egypt.* London: Thames and Hudson, 1992.

Redford, D.B. *Pharaonic Kinglists, Annals and Day-Books: A Contribution to the Study of the Egyptian Sense of History.* Mississauga: Benben Publications, 1986.

Redford, D.B. (ed.). *The Oxford Encyclopedia of Ancient Egypt.* Oxford University Press, 2000.

Sasson, J.M. (ed.). *Civilizations of the Ancient Near East*, 4 Vols. New York: Scribner, 1995.

Stanwick, P.E. *Portraits of the Ptolemies: Greek Kings as Egyptian Pharaohs.* University of Texas Press, 2002.

Thompson, D.J. *Memphis under the Ptolemies.* Princeton University Press, 1988.

Trigger, B.G., Kemp, B.J., O'Connor, D., and Lloyd, A.B. *Ancient Egypt: A Social History.* Cambridge University Press, 1983.

Tyldesley, J.A. *Chronicle of the Queens of Egypt: From Early Dynastic Times to the Death of Cleopatra.* London: Thames & Hudson, 2006.

CHAPTER 2

The Egyptian Land

B EFORE DELVING INTO THE history of the ancient Egyptians, one must know the land to understand its people. The modern name Egypt is derived from the Greek "aigyptose", which probably represents "Hikupta", one of the ancient names of Memphis, the capital of the country during the Old Kingdom.

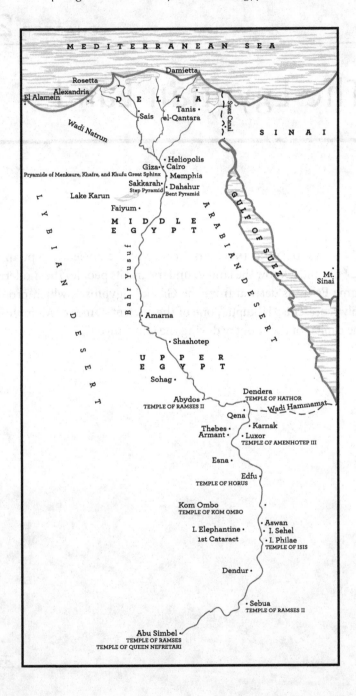

In the old times, the country stretched from the Mediterranean Sea in the north to the First Cataract in the south, a distance of 750 miles by the River Nile. Currently, the lower border is five miles north of Wadi Halfa and the effective land extends to the north of the city Aswan.

Egypt is known as the "Gift of the Nile". On both sides of the Nile, there is desert land. After passing the granite First Cataract, the river passes through a sand storm belt (part of the country where sand storms are more frequent) up to Edfu, after which the valley is composed of limestone deposits as far as Cairo. North of Cairo the river divides into two main branches, namely, the Damietta Branch to the east and the Rosetta Branch to the west, and between them is the Nile Delta. In old antiquity, other branches were mentioned by old writers and by Herodotus. Each year, regularly, the Nile brought not only its waters, the source of biological life, but also its alluvium deposits, which caused and maintained the fertility of the Nile Valley. Based on the course of the Nile, Egypt was divided into regions, namely Upper Egypt in the south and Lower Egypt (the Delta) in the north. In addition to its vital functions, the Nile offered a means of travel and traffic by boats since the early years. It was used by travelers, traders, and government personnel along the valley of the river. During the floods, transport on the river utilized flat-bottomed barges to carry heavy cargo, e.g. building stones and monumental structures from the quarries, to their different destinations. The Nile Valley is the cultivatable region of the country and the habitable part. Other habitable areas were the oases, which could be reached from the valley.

Upper Egypt (Shenau) was administratively divided into 22 zones (nomes in Greek). The 1st was that of Elephantine (Aswan) and the 21st was that of Faiyum, which is a few miles to the west of the valley at Aphroditopolis, taking water from an offshoot from the Nile, called Bahr Youssef. North of Fayum is Birket Qaroon. It was exploited by the Middle Kingdom, who established their capital there.

Lower Egypt had 20 nomes, the first of which was Manupolis.

The deserts on either side of the valley were under government control. The western desert called the Libyan Desert consisted of sand dunes except for the oases: El Dakhla, El Kharga, El Bahrya, and other smaller ones. The Eastern Desert included the Sinai Peninsula and the southern desert of Nubia, although not part of Egypt, was exploited by the Egyptians.

Some of the highlights of the ancient Egyptian civilization included:

1. The development of a reliable system of agriculture and crops.

2. The Nile was the most vital resource for the country.

3. Animal stocks, birds, and fish were plentiful.

4. Papyrus was a source for primitive buildings, making boxes, cords, ropes, mats, and of course material for the scribes. Its flowers were used as decorations.

5. The Nile's mud was used for making bricks and pottery.

6. Limestone, which was found in the cliffs from Cairo to Edfu, was used for construction.

7. Sand bricks were made from the cliffs south of Edfu.

8. The Eastern Desert was very rich in minerals, gold, and precious and semi-precious stones.

9. They made use of red and black granite, alabaster, and basalt.

10. Art and sculpture evolved to a high degree.

11. Construction of exceptional temples, monuments, and pyramids took place.

12. Jewelry and art crafts were superior to those of their neighboring cultures.

13. International trade between Egypt and other countries of the Mediterranean, West Asia, Africa, and the Middle East brought wealth to Egypt.

14. Military adventures, which were mostly successful, also brought in a lot of revenue to the country.

The ancient Egyptians practiced metallurgy by developing tools, farming machinery, chariots, military weapons, and other industrial tools. Jewelry was a profitable and advanced craft.

CHAPTER 3

The Diet, Hygiene, and Social Life

IN ORDER TO STUDY Ancient Egyptian medicine, we must first look at their diet, hygiene, and social life. According to Herodotus (*Histories*, edited by Griffith, 1996):

> Concerning Egypt itself I shall extend my remarks to a great length, because there is no country that possesses so many wonders, nor any that has such a number of works which defy description ... but the people also in most of their manners and customs, exactly reverse the common practice of mankind. The women attend the markets and trade, while the men sit at home at the loom; and here, while the rest of the world works the woof up the warp, the Egyptians work it down; the women likewise carry burthens upon their shoulders, while the men carry them upon their heads. They eat their food out of doors in the streets, but retire for private purposes to their houses ... A woman cannot serve the priestly office either for god or goddess, but men are priest to both; sons need not

support their parents unless they choose but daughters must whether they choose or not ... They are the only people in the world—they at least, and such as have learned the practice from them—who use circumcision. Their men wear two garments apiece, their women but one ... When they write or calculate, instead of going like the Greeks from left to right they move their hands from right to left ... They are religious to excess, far beyond any other race of men and use many ceremonies.

The Nile Valley was fertile and food was plentiful unless there was a drought where the inundation on the banks of the Nile had failed. As Diodorus stated, "The Egyptian physicians looked upon excessive consumption of food as the main cause of disease ... " (1933–1967). Three or four loaves of bread and two jars of beer were considered a good daily ration. The bread of the common people was made of coarse ground cereals. In the Delta, a kind of bread was made from dried lotus.

The following description of an Egyptian's diet is abstracted from Georges Posener's book *Dictionary of Egyptian Civilization* (1959). An Egyptian proverb states: "we are people who do not eat until hungry and when we eat we don't get to be full". Next to bread, fish was probably the main source of protein for the common people. It was consumed in great quantities: raw, roasted, boiled, sun-dried, or pickled. For further reading on the fish in Egypt, see Borgstrom (1961).

A great variety of birds were found along the river and in the marshes, including ducks, quail, and small birds. Hunting with a type of boomerang was an elegant pastime of the rich. The roasted goose was a favorite meal. Cattle were raised at the very beginning of Egyptian history to provide meat, milk, and animal power. Sheep and goats were also used for food. The pig was taboo, and was not eaten, except by the poorer populations; it was considered unhealthy. Slaughtering cattle was done under hygienic rules. Vegetables and fruits were plentiful, particularly, onions, leeks, and garlic; various beans, lentils, melon, watermelons, cucumbers, olives, dates, figs, and grapes were also abundant. All were excellently demonstrated in the reliefs and Stelas. The chief fat used in cooking was olive oil; honey was served for the sweetening of dishes.

As for drink, the Egyptians considered Nile water to be the best on earth; when a princess had married abroad, she had Nile water sent to her from home. Water was carried in goat-skin bags and it was even boiled before use. It was cooled in porous clay jars, which are still used today. Milk was a popular drink in the country. However, beer from barley or other cereals was the popular drink of the rich and the poor. Wine probably only appeared on the table of the rich. The vine has been cultivated in Egypt since the predynastic days and the Egyptian wines were famous until the Islamic period. Drunkenness was known, and moralists raised their voice against the abuse of alcohol and in the Wisdom of Ani, the late didactic book, it is said:

> Take not upon thyself to drink a jug of beer. Thou speakest, and an unintelligible utterance issueth from thy mouth. If thou fallest down and the limbs break, there is none to hold out a hand to thee. Thy companions in drink stand up and say: 'Away with this sot ...' If there then cometh one to seek thee in order to question thee, thou art found lying on the ground, and thou art like a child. (From Papyrus Boulaq 4 of the Cairo Museum [Posener, 1959].)

The Egyptians had three meals a day, while the king and court had five meals a day. All in all, the Egyptian diet, which consisted of coarse bread, milk, vegetables, fish, and occasionally meat, was well balanced and conducive to health. It would lead one to conclude that the dietary causes for arteriosclerosis must have been minimal, except for the rich and royalty. The fish of the Nile is lean and contains low lipids, high unsaturated fatty acids, minerals and vitamins, and even iodine, adenosine triphosphate (ATP), and adenosine diphosphate (ADP). For further information on the dietetics, see Hutchison (1956); for hygiene, see Stetter (1993).

The housing conditions of the poor showed that they liked to sleep out of doors and they used the houses mainly as stores. The rich had big houses divided into bedrooms, living rooms, rooms for servants, stores, granaries, etc. Clerestory windows ventilated the homes. They also had balconies, bathrooms, and sometimes toilets that could be flushed. Sewage was drained into gutters built in the streets. Gardens were essential, rich with ponds and such trees as sycamore, palm, fig, pomegranate, tamarisks, vines, and flowers that were used to decorate tables. The furniture consisted of beds, chairs, mats, baskets, chests, tables, etc. (see King Tut's furniture in the Cairo Museum). They had pet animals such as dogs, cats, ichneumons, or tamed monkeys. People chased flies with fly flaps and protected themselves from mosquitoes with nets. The physicians supervised hygienic practices. Great stress was placed upon the cleanliness of the dwelling, the care of the body, and on dress and diet. Physical exercise – which in the case of the wealthy included swimming – completed an excellent education. Care of the body – skin, hair, eyes, and teeth – was important to the Egyptians in different ways. The Egyptian lady washed her body, shaved it with a bronze razor, used tweezers to remove unwanted hair, anointed her skin, made use of perfumes, painted her lips and cheeks, and took particular care in painting her eyelids. They took great care of their hair according to the prevailing fashion. Cosmetics were widely used (see Manniche, 2006).

The beauty and care of the ancient Egyptian women can be understood from one of the love poems in the Chester Beatty Papyrus I:

The *One*, the sister without peer,
The handsomest of all!
She looks like the rising morning star

At the start of a happy year.
Shining bright, fair of skin,
Lovely the look of her lips,
She has not a word too much
Upright neck, shining breast,
Hair true lapis lazuli;*
Arms surpassing gold,
Fingers like lotus buds.
Heavy thighs, narrow waist,
Her legs parade her beauty;
With graceful step she treads the ground,
Captures my heart by her movements.
She causes all men's necks
To turn about to see her;
Joy has he whom she embraces,
He is like the first of men!
When she steps outside she seems
Like that other *One*!

See also Christiane Desroches Noblecourt's book (1986 and 2001). People married young: girls married at 13 and boys at 15. Marriage between siblings was not common practice, except in the ruling families, since the pharaoh was a descendant of a god they had to "keep the bloodline pure". However, according to Stetter, as late as the second century A.D., two-thirds of the marriages in the city of Arsinoe were brother–sister marriages. Likewise, excessive intercourse was forbidden and stringent penalties were applied to offenses. It was also forbidden during menstruation. Priests were allowed only one wife and monogamy was the rule, while the king and noblemen had their harems. The boys were circumcised between the ages of 6 and 12 (see Stetter); this persisted through Jewish and Islamic religions. There is no doubt about the mixing of the Egyptian and Israelite cultures, since the

---

* Lapis lazuli is a semi-precious rare blue stone.

latter lived in Egypt for more than four centuries before Moses was born in 1350 B.C. The descendants of Joseph and Jacob had lived in Egypt for nearly 400 years. Most of those years had been peaceful. The Israelites lived peacefully, raising their families and tending to their flocks in the fertile Nile Delta region, which the Bible calls Goshen. Adverse conditions started after the Hyksos occupied Egypt and the Pharaohs were being driven out and the oppression started.*

The wealthy Egyptian entertained himself by hunting, fishing, and entertainment at home; they had musicians, dancers, and singers. They also played senet or hounds and jackals, etc. The farmers were, as quoted by Diodorus, "Being relieved of their labors during the entire time of the inundation turn to recreation, feasting all the while, and enjoying without hindrance every device of pleasure". They also had many religious holidays, which they celebrated by worshiping and making offerings. These ceremonies were important to their spirituality, not only for their way of life on earth, but also for their afterlife in the netherworld.

It is clear that the ancient Egyptians did not smoke, had no syphilis, and very rarely tuberculosis; diabetes and hypertension were uncommon. Their diet and hygienic conditions were as Diodorus said, "The whole manner of life was so evenly ordered that it would appear as though it had been arranged according to the rules of health by a learned physician, rather than by a lawgiver" (1933–1967). They had entertainment and rest after work, and the religious beliefs resulted in internal contentment. They lived in a good climate with little change that allowed them to enjoy the sun all year. The main causes of death were different from what we see in modern times. Common causes of death were infections during epidemics, intoxication, parasitic diseases, poisoning, war, accidents, and forced labor.

---

* For more details about the Israelites in Egypt, see the five books of Moses; the Old Testament.

From Stetter, we know that the ancient Egyptians wished for a life filled with vitality, freshness, and good health; death was just as vital. Good health meant wellbeing (snb). Common well wishes were: good health, wellbeing, and living well; may the heart be happy and all parts of the body well functioning.

Before discussing embalming, it is appropriate to briefly describe how they viewed death. The ancient Egyptians made elaborate preparations for death. To them, it was not the end.

SUGGESTED READINGS ON DIET, HYGIENE, AND SOCIAL LIFE

Brier, B. and Hoyt, H. *Daily Life of the Ancient Egyptians*. New York: Greenwood Press, 1999.

Darby, W.J. (ed.). *Food: The Gift of Osiris*. University of Michigan Academic Press, 1977.

Desroches-Noblecourt, C. *La femme au temps des pharaons*, Ed. Stock, 1986 and 2001.

Engelmann, H. and Hallof, J. Der Sachmetpriester, ein früher Repräsentant der Hygiene und des Seuchenschutzes. *Studien zur Altägyptischen Kultur*. 1996; 23:103–46.

Hawass, Z. *Silent Images: Women in Pharaonic Egypt*. Danbury, CT: Harry N. Abrams, Incorporated, 2000.

Manniche, L. *An Ancient Egyptian Herbal*. British Museum Press (Distribution); 1st University of Texas Press edition, December 31, 2006.

Robins, G. *Women in Ancient Egypt*. Harvard University Press, 1993.

Wilson, H. *Egyptian Food and Drink*. Shire Publications, 1997.

CHAPTER 4

The Afterlife in Ancient Egyptian Beliefs

THE BELIEFS OF THE ancient Egyptians about the afterlife were richly developed with the inception of the concept of the soul in the religion of the Old Kingdom. The cult of Osiris envisaged a posthumous existence, which could be enjoyed not only by the royals, but also by all people of different ranks.

Osiris, the King of the Dead and also a dead king, presented the hope of survival in the afterlife and the deceased could be identified with Osiris. To attain this resurrection, the deceased had to be tested in the presence of 42 assessor-gods. The Egyptian's aspiration for his existence after death was that he should have access to the best available to him in his life on Earth. For life hereafter, it was necessary that his name should continue to be the same, and that his body should remain intact. Also it was essential that he should be supplied with the necessary food and drink and other desirable objects.

The Egyptian believed that he had a spiritual entity called the *ka*, which, to a certain extent, corresponds with the "self". It was born with him, formed an integral part of his being, and was distinct from the bodily "self"; for which one provided all the paraphernalia of funerary equipment, food, and drink in the tomb. The tomb itself was known as the "House of the *ka*".

Another spiritual entity was the *ba* or *bai*, which was supposed to leave the body after death and is represented by a human-headed bird; it helps man assume different forms in life and death.

In the "Judgment Hall", Ani, the scribe of the *Book of the Dead*, and his wife, Tutu, stand by the scale in which his heart, which for the Egyptian was the center of intelligence, is being weighed. The feather, which represents truth, is weighed and watched on a scale by Anubis, the jackal god of Necropolis. Thot, the ibis-headed god, is recording each movement. The deceased denies a specific sin to each assessor; the sins related to behavior in life and reveal the general moral attitude of the Egyptian. At the conclusion of his negative confession, the deceased has declared truth or voice and is led to the presence of Osiris (see Figure 4.1).

METHODS OF EMBALMING

The preservation of the body was important for the resurrection of the dead in the afterlife. Embalming techniques were different in different dynasties and are described by Herodotus (Book II, chapters 86–89) (*Histories*, edited by Griffith, 1996) and abstracted here:

> There are certain individuals appointed for the purpose (embalming) and who profess that art, these persons after anybody is brought to them, show the bearers some good models of corpses ... having settled upon the price, the relations immediately depart and the embalmers, remaining home. There was a special place for mummification to which dead bodies were taken to perform the embalming in the most costly manner. First, with a

FIGURE 4.1   The Weighing of the Heart from the *Book of the Dead* of Ani. At left, Ani and his wife Tutu enter the assemblage of gods. At the center, Anubis weighs Ani's heart against the feather of Maat, observed by the goddesses Renenutet and Meshkenet, the god Shay, and Ani's own *ba*. At the right, the monster Ammut, who will devour Ani's soul if he is unworthy, awaits the verdict, while the god Thoth prepares to record it. At the top are gods acting as judges: Hu and Sia, Hathor, Horus, Isis and Nephthys, Nut, Geb, Tefnut, Shu, Atum, and Ra-Horakhty (artwork created c. 1300 B.C.). Photographed by the British Museum, published 2001 (commons.wikimedia.org).

crooked piece of iron they pull out the brain by the nostrils, a part of it they extract in this manner, the rest by pouring in certain drugs. Next, they make an incision in the flank with a sharp Egyptian stone, like a knife. Then they empty the whole of the inside, and they clean the cavity, and rinsing it with palm wine, scour it out again with pounded aromatics: then having filled up the belly with pure myrrh pounded, and cinnamon and all other perfumes, frankincense accepted. They sew it up again then they steep the body in natron, keeping it covered for 70 days. This period included the wrapping as well, and it was unlawful to leave the body longer in the brine. When the 70 days passed, they washed the corps and then wrapped the whole body in bandages cut out of flax cloth, which they smear with gum, a substance the Egyptians used instead of paste.

The relatives then received the body and placed the mummy in a wooden case in the shape of the body. These cases were kept in a sepulchral repository, where they stand it upright against the wall. This embalming process was the most costly method to prepare the dead. For such as to choose the middle method, they prepare the body thus: they first fill syringes with cedar oil, which they inject into the belly of the deceased, without making any incision or emptying the inside, but sending it up through the anus. They then close the aperture to hinder the injection from flowing backward and lay the body in brine for the specified number of days, on the last of which they take out the cedar oil that they have previously injected. Such is the strength of the oil that it brings away with it the bowels and the inside in a state of dissolution. On the other hand, the natrium dissolves the flesh, so that, in fact, there remains nothing, but the skin and the bones; when having so done, they give back the body without performing any further operation upon it.

The third method of embalming, which is used for such people who have but scanty resources, and it is done as follows: after washing the inside with syrmaea, they salt the body for 70 days, and return it to be taken back. The deceased wives of important men were not given the same quality and are not given to be embalmed immediately after their death, neither are those that may have been extremely beautiful, or much celebrated; however, they deliver them to embalmers after having been three or four days deceased.

Herodotus died in 406 B.C. and, therefore, his description may not represent accurately the practice of embalming performed a thousand years before him.

According to Elliot Smith (1908), the embalming incision usually caused a large, vertical fusiform, gaping wound in the left lumbar region, extending from the iliac crest, about two to three 3 inches behind the anterior superior spine, to the costal margin. It may be extended forward or extend lower down in front of the iliac spine. As a rule, no attempt was made to close the wound, which was then covered with a plate usually of wax; however, in royalty, the wax was concealed by a plate of metal and/or gold on which the *wedjat*-eye was engraved; it was a sign of protection and healing. (For further details on embalming practices, see Ziskind and Halioua's chapter on mummification.) The body cavity having been opened, the intestines, liver, spleen, kidneys, stomach, and pelvic viscera and most of the vessels were completely removed. The diaphragm having been cut through, the lungs were freed by severing the bronchi; or, in some cases, the lower end of the trachea. The heart was left in the body but never exactly in the normal position. Generally, it was pushed upwards into the upper part of the right side of the thorax; sometimes, it was left in the middle line in front of the vertebral column or again it is found in the left side of the chest. Sometimes, only the arch and a small part of the aorta were left behind. Occasionally, the whole aorta and iliac vessels were left behind. The body and the organs were put in saline. The hearts left in the body were always well preserved. In many cases, the valves are intact and it is often possible to recognize the chordae tendineae and muscli papillares. As a rule, the organ is considerably damaged by the operator, but not intentionally, especially the auricles, and, in others, the ventricles. The cavities of the heart were in many subjects tightly stuffed with mud or a mixture of mud and sawdust. The viscera, after having been removed from the salt bath, were thickly sprinkled with coarse sawdust of various aromatic woods, and when still flexible were molded into shape and wrapped in linen. The small intestines were usually bent upon themselves to form an elongated parcel of parallel bands. Among these bands was placed (when the viscus was still flexible) a wax image of one of the four genii, usually the hawk-headed Qebehsenuef. Then, after being sprinkled with sawdust, the mass was wrapped in linen bandages. In the case of the liver, a wax statuette of the human-headed Amset was inserted in most cases. These viscera were found in the body cavity usually in specific positions: the intestines on the right side of the abdomen vertically and the liver transversely in the lower part of the chest. After the various parcels of viscera had been returned to the body and had been tightly packed with sawdust or coarser fragments of wood, the relatively empty pelvis was also stuffed with sawdust.

Flowers and vegetable substances, especially onions, were often found among the wrappings on the surface of the mummy or inside it.

An analysis of the mineral salt (natron) showed it to be the natural soda found in Egypt chiefly in Wady El-Natroun and it is a mixture of sodium carbonate, sodium chloride, and sodium sulfate. It contains also a certain amount of clay and calcium carbonate. After 1567 B.C. or the beginning of the new kingdom, the heart was excised for fear of testifying against the deceased. The viscera were mixed with myrrh and soda and kept in ornamented urns (the canopic jars), which were placed in the tomb with the mummy.

Professor H.E. Derry with Engelbach (1942) have studied the evolution of embalming methods (cf. *Introduction to Egyptian Archaeology* by Engelbach [1946]). They discovered that in the Old Kingdom, the organs were taken out and preserved in canopic jars. The mummy consisted, therefore, of bones and shreds of tissue, as they did not use preservation. Starting with Dynasty XI, they used preservation measures such as oil, natron, resin, beeswax, and sand. In Dynasty XII, they started wrapping the body carefully with linen soaked in resin. Starting in Dynasty XXI, the viscera were returned into the body with their protective deities. These were: the human-headed god (Imseti) for the liver, baboon-headed god (Hapi) for the lungs, the wolf-headed god (Duamiutef) for the stomach and small intestine, and the hawk-headed god (Kebehsenuaf) for the large intestine. There were no canopic jars with the priests of Amun, but the royalties of Tanis had them even without viscera.

Buckley and Evershed have reported on "Organic Chemistry of Embalming Agents in Pharaonic and Greco-Roman Mummies" (2001). They used modern investigative techniques. Their results showed that a mixture of substances was used, namely, dried oils, coniferous resin, and beeswax. The latter constituted a greater portion of the organic material compared to the resins. They added: "We do not know all there is to know about Egyptian mummification and caution needs to be exercised when making assumptions about the materials that the ancient Egyptian embalmers might have used".

In the 17th century A.D., an unthinkable practice developed, namely, the use of the mummy tissues as medicine. Pieces of the mummies were bought at high prices and recommended by physicians of that time to treat epilepsy, gout, and other ailments. Even the French King, Francis I, thought it was a panacea (cf. Estes, 1983). Mummy tissues were also applied to wounds in Edinburgh.

SUGGESTED READINGS ON METHODS OF EMBALMING

Adams, B. *Egyptian Mummies*. Shire Egyptology 1. Aylesbury: Shire Publications, 1984.

Andrews, C. *Egyptian Mummies*. Harvard University Press, 1984.

Balout, L. (ed.). *La momie de Ramses II: Contribution scientifique à l'Égyptologie sous la direction de Lionel Balout et C. Roubet, avec la participation de Ch. Desroches-Noblecourt, etc.* Paris: Éditions Recherche sur les Civilisations, 1985.

D'Auria, S. Mummification in Ancient Egypt. In *Mummies and Magic: The Funerary Arts of Ancient Egypt*, ed. S. D'Auria, P. Lacovara, and C.H. Roehrig, 14–19. Museum of Fine Arts, Boston, 1988.

Ikram, S. and Dodson, A. *The Mummy in Ancient Egypt: Equipping the Dead for Eternity*. London: Thames and Hudson Ltd, 1998.

Iskander, Z. Mummification in Ancient Egypt: Development, History and Techniques. In *An X-Ray Atlas of the Royal Mummies*, ed. J.E. Harris and E.F. Wente, 1–51. University of Chicago Press, 1980.

CHAPTER 5

Medicine, Religion, and Magic

F ROM THE EARLIEST TIMES, medicine, magic, and religion were associated with each other and confused. The physician might with his magic exert a favorable influence over ghosts and devils; with his medical knowledge, he could give the patient his remedies; and with his correct ritual, he might placate the gods to free his patient from illness. There was a logical connection to the medicine man between these three elements. In ancient Egypt, the physician was a priest, a magician, and a doctor. This may explain why later prophets and saints were considered miracle performers and healers. After the Egyptians, physicians combined medicine with philosophy, astronomy, mathematics, metaphysics, theology, and even politics and art (cf. Clendening, 1991; Monro, 1951; Moon, 1909; Arberry, 1950). In many cultures, e.g. the Hebrews, the Greco-Romans, the Arabs, the Medieval Europeans, and even more modern nations, there were famous physicians distinguished in philosophy, poetry, anthropology, etc. A few prominent examples from history include Imhotep, Hippocrates, Galen, Avicenna, and Maimonides. The early medicine man was considered a superman whose function was to control the dark powers that threatened human life both seen and unseen. We have learned from the ancient Egyptian, ancient Greek, Hebrew, and Arab/Muslim philosophers that man has three sides to his nature, namely body or physic, mind or psyche, and spirit or pneuma. Modern medicine deals with the first two; and there are also spiritual healers practicing holistic medicine. Modern Western-trained physicians are expected to treat their patients based on scientific evidence. This includes the appropriate use of psychotherapy in the treatment of psychological ailments. The links between the physical and psychological have been confirmed. Recently, it has been shown that stress can alter immunological, neurochemical, and endocrine functions, thus increasing the risk of cancer, for example. (For more details on this subject, see Thacker et al., 2006.) Even in modern times, physicians are often expected to have the attributes of scientists, healers, and priests.

One of the earliest panegyrics of the healing art is found in the *Apocryphal Book of Ecclesiasticus* compiled by a Jerusalem Notable, namely Joshua Ben Sirach, in the second century B.C. It states:

> Honor the physician with the honor due unto him. For of the most High cometh healing, and he shall receive honor of the King. The skill of the physician shall lift up his head, in the sight of great men he shall be in admiration ... Then give place to the physician, for the Lord hath created him; let him not go from Thee, for Thou hast need of him. There is time when in their hands there is good success.

In the Ebers Papyrus (1875), the physician believed "The charms have great power over the remedy". Also, Ambroise Paré (1598), the famous French surgeon of the 16th century said: "Je le peansais, Dieu le guearit", i.e. I dress the wound and God heals it. However, in our modern times, since the industrial revolution, these ideals are more or less lacking and medicine is viewed as an industry (i.e. healthcare industry).

MEDICINE AND RELIGION IN ANCIENT EGYPT

According to Herodotus, "the Egyptians are very religious to excess, far beyond any other race of man". Egyptian medicine, like most of her cultures, began with the priests. While we will not describe in detail the Egyptian theology (cf. Erman, 1937; Budge), a general idea will be given as it relates to medical science. The Egyptians were polytheists; they had numerous gods and goddesses, which they could picture in the form of a man, a woman, an animal (zoolatry), or a plant. The ancient Egyptian had to look to things about him, the sky above him, the environment around him, and the earth beneath his feet to find explanations to life and death events. For example, Hathor, which means the house of Horus. Hathor was one of the oldest goddesses of Egypt representing the great mother or cosmic goddess who conceived, brought forth, and maintained all life. She not only nourished the living with her milk, but was also said to supply celestial food to the dead in the underworld. (For more details, see *Who's Who in Egyptian Mythology*.) Very early in their history, they saw in the heavens the outline of a vast cow, which stood athwart, the vault with its head in the west, the earth lying between the four feet, while its belly, studded with stars, was the arch of the heavens. The four legs represented the pillars supporting the sky. In another locality, a woman's outline replaced the cow. To others, the sky was a sea supported on four pillars.

They imagined the sun as a child or cow or woman arising every morning and sailing every evening to the west. They also compared it to a flying hawk taking its daily flight across the heavens and the sun-disc with its spreading rays or wings became one of the most common symbols of their religion. The Nile, to them, seemed to arise from a source beneath the earth and it ended in a sort of ocean and from this, the sun-god developed. It was considered the center of life where the world began and was worshiped as a god, Hapi. From himself, he begat four children, viz. Shu and Tefmit, who represent the atmosphere, Keb the earth, and Nut the sky. Osiris and Isis, Seth, and Nephthys, together formed with their primeval father, the sun-god, a circle of nine deities, which is the "ennead" of which many temples later possessed a local form. This correlation of father, mother, and son strongly influenced the theology of later periods so that each temple had an artificially created triad (cf. Frazer, 1936). Other local versions of the myths of the origin of the world were circulated. One of them described Re as the King, who ruled over people and the Earth. The people plotted against him, so Re sent a goddess, Hathor, to destroy them, but she was diverted after she had destroyed them in part. The cow of the sky then raised Re upon her back so that he might forsake the ungrateful Earth and dwell in heaven. There were also gods of the netherworld. Osiris was the king of the dead, who succeeded the sun-god as king on the earth, aided in his government by his sister-wife Isis. Anubis was the god who was concerned with embalming. Osiris was slain by his brother Seth. Isis gave birth to Horus who defeated Seth in revenge for his father and assumed the earthly throne of his father. Seth attacked him, but Thoth, the god of letters, defended him. Some of these gods were only mythological figures, but others in time became the great gods of Egypt. These beliefs of the ancient Egyptians are considered to represent spiritual events in the creation of the world (Sun – Moon – Earth – animals – plants). Re had almost universal worship and the main center of his cult was at On (Heliopolis of the Greeks) with a symbol of an obelisk, although, at the Edfu temple, he appeared as a hawk under the name Horus. The moon, as the measurer of time furnished the god of reckoning, of letters, and of wisdom (Thot). The Greek Hermes, who lived at Shmûm or Hermopolis, was identified with Thot. The sky was at first called Nut, then a cow-goddess Hathor, then in the form of a cat (Bast), and then as a lioness (goddess of storms and terror). Ptah was the god of art and the patron of artisans in Memphis. The Egyptians saw the manifestations of their gods in the animals, which surrounded them.

Animal worship only occurred during the decline of the culture. Animals were petted in the temples, but not actually worshipped.

Various sources reveal the complexity of the representations and relationships between gods and goddesses. Thot was a human deity and also god of wisdom and souls. He was often represented as a man with the head of an ibis and sometimes represented as a whole ibis bird and sometimes as a baboon. Horus was represented as part falcon, not to be confused with Montu, the god of wars. Sometimes, the goddess Hathor is mentioned to be his mother; in others, it is Isis. It may be assumed that one deity is an incarnation of another. Bastet, the mother goddess, is embodied as a cat. Sekhmet, an aggressive goddess, has a lioness face. Gods were male and female and animals or combined human and animal representations, e.g. Sekhmet, Kheprer or Kheprera – the god who represented the rising of the sun was associated with the scarab of the sacred beetle. Like the rising sun, he was self-created. He was portrayed as a beetle-headed man, which was a symbol of resurrection and fertility, and it was worshiped in early Egyptian history, associated with Ra. The beetle amulets were buried with the cadaver to help its resurrection. The goddess of fertility was a snake – Tawaret (goddess of childbirth) – and was represented as a hippopotamus with features of a crocodile and lion to keep evil from the mother and child. Ammit, the goddess who devours the hearts of sinners and denies them life after death, was partly crocodile, panther, lion, and hippopotamus. (For more details, see *The Deities tables in Ancient Egypt* by Oakes and Gahlin, pp. 274–275). See also: *Who's Who in Egyptian Mythology* (Mercatante, 1978).

In their temples, the Egyptians used to have many ceremonies. People gathered on feast days to share the generous offerings and frequently for commemoration of the gods. With the development of the nation, the Pharaoh was the chief and sole official servant of the gods, thus starting the history of a state form of religion. The Pharaoh himself worshipped the god and goddess. He was represented in each temple by a high priest by whom all offerings were presented and the temple administrated. He was assisted by a group of priests, who were actually laymen, and priestesses who danced and sang for the god on festive occasions. The priests were professional servants of the god. There was the belief in the eternal sojourn after death to which they devoted so much to the building of eternal houses. The Egyptians believed that the body was animated by a vital force, which they imagined to be a counterpart of the body, which came into the world with it, passed through life in its company, and accompanied it into the next world. This was called *ka*. In addition, every person possessed a soul. Further elements of personality seemed to them as present as the shadow. This is compared now with what is called the metaphysical or astral and the etheric bodies of the spiritualist. It was the basis of the "Temple Sleep" therapy by ancient Egyptians, a form of spiritual therapy (46) similar to the present-day yoga and hypnosis. They believed that the world of the dead was in the west where the Sun descended into the grave every night, and that is why they built their cemeteries in the west of the country (e.g. the Pyramids). The dead had to await the return of the solar barque that they may bathe in the radiance of the sun-god and that the bow-rope of his craft may draw them with rejoicing through the long caverns of their dark abode. The sun-god would receive only those who have done no evil. This was the first ethical test that makes the next life dependent on the character displayed on earth. They imagined also how the dead would be received by Re and the servants of God. Because of this, they contemplated on death without dismay. They said of the dead, "They depart not as those who are dead, but they depart as those who are living". They will sit on the throne of Osiris and they will be glorified. In spite of this departure, the Egyptians were never able to separate the afterlife from the body and could not imagine the resurrection of the deceased without it (see the *Book of the Dead*). Therefore, they paid much attention to the repository of the dead by building big tombs, *mastabas*, and the Pyramids. Ptah was the god of architects and craftsmen and he became the supreme mind. They believed that the entire universe was a creation of his mind. It was believed that gods and people's actions resulted from the mind of the god working in them.

As the empire evolved, the theory that god is universal started to appear. This universal god was referred to as Amon at Thebes and as Ptah in Memphis and as Re in Heliopolis. Under the reign of Amenhotep III, Re was called Aton (an old name for the material sun). Akhenaton then replaced Aton (spirit of Aton) and a city was built for his residence called horizon of Aton, "Akhetaton", now known as Tell El-Amarna. Akhenaton, then a young Pharaoh, embraced the concept of one powerful god who created all of nature. He believed in the creator's beneficial purpose for all the creatures, even the meanest. For the birds fluttering about in the lily-grown Nile marshes, to him seemed to be uplifting their wings in adoration of their creator and even the fish in the stream leaped up in praise of God. He based the universal sway of God upon his fatherly care of all men alike irrespective of race and nationality. To the proud and exclusive Egyptian, he pointed to the all-embracing bounty of the common father of humanity, even placing Syria and Nubia before Egypt in his enumeration. In this respect, Akhenaton was the first prophet of the Egyptian monotheism (see one of his hymns).

A little more detail about Akhenaton is provided here. Amenhotep III delegated his son Amenhotep IV to be a caregiver due to his ill health. When Amenhotep III died, his son took over the throne. He changed his name to Akhen-Aton (means beneficial to Aton), i.e. the god expressed in solar discs. Akhenaton had the belief that there is only one god and that the other gods in ancient Egypt's history for thousands of years were not true gods. He was the first monotheist, long before Moses. He became very spiritual and mystic. He left his capital in Thebes and built another capital at Tell El-Amarna as a beautiful city. He had physical deformities, namely a very elongated face, much exaggerated chin, almond-shaped eyes, wide lips, and gynecomastia. However, he was married to the most beautiful queen Nefertiti, the icon of Egyptian beauty. He was dedicated to his new religion at the expense of the

FIGURE 5.1   Akhenaton (L) and Nefertiti with Three Daughters (commons.wikimedia.org).

government and the army. He died in the 17th year of his reign. (For more details, see Weigall; see also Figure 5.1.)

In connection with religion, the highly developed hygiene of the Egyptians even overshadowed their medical knowledge. It had a powerful influence compared with the social hygiene of Judah. Herodotus (30), in his second book, part 37, wrote about the Egyptian priests:

> They are religious to excess, far beyond any other race of men, and use the following customs: They drink out of brazen cups which they scour every day, there is no exception to this practice. They wear linen garments, which they are especially careful to have always fresh washed. They practice circumcision for the sake of cleanliness, considering it better to be cleanly than comely. The priests shave their whole body every other day, so that no lice or other impure things may adhere to them when they are engaged in the service of the gods. Their dress is entirely of linen and their shoes of the papyrus plant; it is not lawful for them to wear either dress or shoes of any other material. They bathe twice every day in cold water and twice each

night, besides which they observe, so to speak, thousands of ceremonies. They enjoy, however, not a few advantages. They consume none, except of their own property, and ate at no expense of anybody. Everyday bread is baked for them of sacred corn and a plentiful supply of beef and of goose's flesh is assigned to each, and also a portion of wine made from the grape. Fish they are not allowed to eat and beans – which none of the Egyptians ever sow, or eat, if they come up of their own accord, either raw or boiled – the priests will not even endure to look on, since they consider it an unclean kind of pulse. A single priest for each god has the attendance of a college, at the head of it is a chief priest; when one of these dies, his son is appointed in his own place. (For more details on hygiene see Stetter.)

These principles are similar to Jewish medicine, where the preservation of the physical wellbeing was looked upon as a religious command (Chopra, 1990). "And live through them, but not die through them" (Yoma 85b, based on Lev. xviii–5). The neglect of one's health was regarded as a sin. Purity, which is the aim of most of the Biblical sanitary laws, was to be not only physical, but also moral and religious. Islam also commands: Cleanliness is a command of faith (meaning physical and moral). Similar commands are found in Christianity, Buddhism, Zoroastrianism, etc.

It is also known that priests were exempted from taxes when the Pharaoh appointed Joseph to be in charge of the land of Egypt. The Egyptians paid for the corn, but when they could not pay, they gave free labor to the state. The land of the priests was exempt from the tax of the one-fifth part of the produce, which was not levied. This shows the highly esteemed position of priesthood and religion in ancient Egypt.

The shrines for worship found in Egypt are the temples. In the predynastic period, the temples were temporary for worship. In the late predynastic period, the first permanent worship site was the center of Hierakonpolis and the early dynastic shrine was at Elephantine by the first cataract.

Two types of temples were found:

1. Mortuary temples where rituals connected with deceased kings were performed.

2. Cult temples erected in honor of a particular god. Few are for combined purposes like that of Seti (ca. 1294–1279 B.C.) in Abydos.

Most temples seen now are relics of the New Kingdom.

To name a few:

- Satet at Elephantine

- Temple of Luxor

- The temples of Karnak

- Temples of Aton (Akhen-Aton)

- Temple of Hathor

- Temple at Dondera

- Temple at Philae

They all show superb architecture, art, and rituals (for more details, see Oakes and Gahlin). Very few temples were discovered to be from the Ptolemaic era. During the Christian era, some ancient temples were converted to churches. The Isis temple in Philae was converted to the Virgin Mary Church. In Aswan, Jews settled and built a temple of their own (Oakes and Gahlin).

From the earlier-mentioned brief descriptions, it is evident that the ancient Egyptians worshiped a large number of gods and goddesses. The origin of the deities is not well documented. Their names were initiated by the priesthood, and they were changed from time to time, resulting in fights amongst the theological entities. This set the stage for the arrival of Moses who declared faith in one God as being the creator of the universe, the omnipotent, omnipresent, and omniscient. (See the book of Exodus for more details on Moses.)

MAGIC AND MEDICINE IN PHARAONIC EGYPT

Professor Dr Paul Ghalioungui studied this in his book *Magic and Medical Science in Ancient Egypt* (1963). Only a short review is presented here:

> Magic and medicine were closely allied in the time of the Pharaohs. Magic is generally regarded as the older of the two and never has lost hold on its offspring. Doing some benefit by mysterious or miraculous powers reflected the belief in the power of magicians. The Egyptians believed that wizardry could achieve results, which were not possible to achieve by other means. It is difficult to separate magic from religion, since magic was a form of applied religion (*religio privata*). Magic was not only used to treat disease but to solve other problems of daily life. It was practiced not only for therapy but also for prophylaxis, sometimes benevolent to some and malevolent to others. Practitioners of Magic had their instruments and objects ready to be used when called. Studies of Egyptian therapeutics are often discounted when the ever-present accompaniment of magical formulas, amulets, and incantations are taken into consideration. Disease was generally not attributed to a disturbance of normal functions, but to a malign spirit or god who had entered the body and attacked it (comparable to bacteria, which are invisible to the naked eye, in modern medicine). Unless this evil spirit could be expelled, recovery was not expected. The evil spirits or demons insinuated themselves, into the individual by the mouth or nostrils (compared with

droplet and air-borne infection of modern medicine) or the ears, and once inside they devour his vital substance. The disease in their view depended on the progress made by the destructive power of the invading spirit, its virulence at times and seasons of the year and on lucky and unlucky days. The physician-priest magician had to discover (diagnose) the nature and if possible the name of the evil spirit and then to expel it by all means.

Incantations were recited during the preparations of some remedies, which provided them with their efficacy. One of these was:

That Isis might make him free, make him free. That Isis might make Horus free from all evil that his brother Seth had done to him when he slew his father Osiris. 'O Isis! Great enchantress, free me, release me from all evil, red things, from the fever of the god and the fever of the goddess, from death, and death from pain and the pain which comes over me, as thou hast freed, as thou hast released thy son Horus, whilst I enter into the fire and go for the water.

While the patient swallows his remedy, another incantation was recited, viz. "Welcome remedy, welcome, which destroyest the trouble in this my heart and in these my limbs. The magic of Horus is victorious in the remedy".

Similar incantations are still used in contemporary Egypt. The Coptic priests and the Muslim clerics still use them in the name of Christ or God. In the countryside, especially, variations of such incantations are still heard.

Similarly, incantations were written on a papyrus and washed off into the medicine to be swallowed by the patient. This is still present in Eastern countries including Egypt.

Spells had to be said by the magician or priest in his name or a god's name. They could be done in the form of commandments, threats, or oppositions, or by uttering certain names or assertions of immunity.

Amulets were used frequently and are still used by older generations of Egyptians; some are called "Hegab" (Erman, 1937; Estes, 1989; Brovarski et al., 1982). Charms and knots were made of knotted threads, bits of cloth, etc. Arabic and Hebrew incantations, with numerical tables, are used to drive out evil spirits (see Shams El-Maarif [*Arabic*]). The use or administering of strange substances, such as hairs of various animals, dung of the hippopotamus or crocodile, different plants, animal matter, and minerals, was often cited in the incantations.

Magic ceremonies were resorted to when the physician's skill was baffled. These conditions may have been psychosomatic disorders, which were treated at famous shrines.

The priests, asking the gods to intercede by prayers to cure the patient or sometimes to threaten the god if he disobeyed the injunctions, practiced sacerdotal methods. These are also included in the temple treatments.

In the available ancient documents, there is quite a rich magic pharmacopoeia. The best practitioners were rationalistic and could carefully diagnose cases for whom magic was most suited, those for whom incantations accompanied by drugs were most suited, and those for whom incubation in the temple promised the best result. In the early days, superstitions and sacerdotal measures probably dominated and this influenced the Egyptian medical history even in the educated class, but gradually actual drug therapy came to have more influence.

Even now, some villages in Egypt still believe in the role of evil spirits in the etiology of diseases. They try to chase it out by incantations, spells, etc., and by a big party using music, songs, dancing, and loud drums (called *Zarr*).

Sorcery performances are different from sacerdotal measures. The former were done for entertainment. The story of Aaron and Moses' rod turning into a snake, which engulfed the snakes of the sorcerers by divine power, is well known in the Torah and the

Qur'an (41). For more information, please see the Old Testament, Exodus 4:2–4 and 7:7–12.

For more extensive information on magic in ancient Egypt, see the following reference:

James, T.H.G. *An Introduction to Ancient Egypt*. British Museum Publications Limited: London, 1979, pp. 128–154.

CHAPTER 6

Egyptian Medicine

A NCIENT EGYPTIAN MEDICINE HAS received scarce attention by medical historians. In most of the standard texts about the history of medicine, accounts of Egyptian medicine are not found. Most of these references place enormous emphasis on Greek medical knowledge and practice. One of the reasons for this neglect is that many medical historians have been unaware of the original ancient Egyptian sources. Moreover, the nature of Egyptian magic, which is the parent of medicine, was misunderstood so as to give a false impression of Egyptian medicine. To handle this subject adequately, one must get the aid of an Egyptologist. There are several of those such as Professor Hermann Grapow of Berlin and his school. Many publications and books came out in the 20th century and more of them recently. Here are a few relevant authors: Ghaliongui, Nunn, Estes, Horbiana and Ziskind, Riad, Fyad, Manniche, Moursi, Stetter, and Leca. To quote an older example, Dr. Paul Cardano, who was killed in France in the war of 1914, wrote to his teacher saying:

> At the time you could still teach us that the history of medicine began with the early Greeks and that Greece was the wellspring of all medical practice and medical

thinking! Now I know that doctors and medical thinking existed thousands of years before the first Greek physician appeared on the scene. Still older chapters of medical history may some day be brought to light. If I am allowed to survive this war, I shall devote myself to those chapters in Egypt.

In Egypt, there are medical Egyptologists, e.g. H. Kamal, G. Sobhi, P. Ghalioungui, Fayyad, W. El-Sisi, S.M. Arab, and M.K. Hussein.

In describing ancient Egyptian medicine, Warren R. Dawson (1953) said:

The wisdom of the Egyptians is indeed proverbial, and although they were incapable of that characteristic of the Greek mind—true philosophy and abstract thought—there is no doubt that they were a very highly gifted people, with a great capacity for practical achievement. There can no longer be any reasonable doubt that the foundations of medical science were laid in Egypt more than fifty centuries ago ... and indeed a nation which had acquired sufficient knowledge and skill to plan and carry out feats of engineering and architecture, such as the pyramids as early as the fourth millennium before Christ, and whose mathematical knowledge, whilst wholly practical in aim included highly complex calculations involving the principles of angles, cubic capacity, the square root and fractional notations was obviously far ahead of its contemporaries in intellectual capacity.

(SEE OAKES AND GAHLIN, P. 50)

He concludes:

From Egypt we have the earliest medical books, the first observations in human and comparative anatomy, the first experiments in surgery and pharmacy, the first use of splints, bandages, compresses, and other appliances, and the first anatomical and medical vocabulary and that is an extensive one. Surgical techniques were first produced. (See Nunn pp. 163–83.) In general terms, it may be said that the popular medicine of almost every country in Europe and the Near East largely owes its origin to Egypt and in its various migrations it has preserved its ancestral form and its very words and phrases almost intact throughout the ages.

(SEE RIAD, P. 41)

Not only were many well-known drugs of universal vogue first used by the Egyptians, but also their properties and traditions, which occur in the works of Pliny, Dioscorides, and Galen, and even in the Hippocratic collection itself, were clearly borrowed from Egypt (53). These later writers, and others who followed them, are the sources from which the compilers of herbals and books of popular medicine mainly drew their materials from and the works of classical writers are often merely the stepping stones by which much of the ancient medical lore of Egypt reached Europe (Singer, 1962). Early medical books in Arabic, Hebrew, and other Semitic languages have also drawn largely from the same sources and subsequently from Greek medicine (Edelstein et al., 1967).

In the ancient world, Egypt was the pioneer country as regards the evolution of medicine. The writers of antiquity were unanimous in praising the skill of the Egyptian physicians. Homer (1102 B.C.?) says in the *Odyssey*: "In Egypt the men are more skilled in medicine than any of the human kind" (Homer, 2003). Herodotus (ca. 484–425 B.C.) relates that both the Persian kings, Cyrus and Darius, had Egyptian physicians. He also notes the specialization of Egyptian medicine and he remarks:

Each physician is a physician of one disease and no more; and the whole country is full of physicians, for some profess themselves to be physicians of the eyes, others of the head, others of the teeth, others of the affections of the stomach and others of the more obscure ailments.

(ABSTRACTED FROM GHALIOUNGUI'S *THE PHYSICIANS OF PHARAONIC EGYPT*, 1983)

Regarding women's care, there were Egyptian women practicing midwifery as attested in the Westcar Papyrus (collection of the National Museum in Berlin from the Middle Kingdom). The Erment Stone depicts midwives helping Cleopatra give birth to Caesarion. Ghalioungui has proved that women, at least in the Old Kingdom, had access to the medical profession. Both in diagnosis and in therapeutics, astonishing progress was made. Even pathology attracted widespread interest, and according to the Edwin Smith Papyrus, dissection of human bodies was apparently practiced. The medical schools of Egypt, closely associated with temples, were highly esteemed. Gods were attributed miraculous powers in healing the sick and restoring apparently dying persons to health. The ancient Egyptians had numerous deities to whom were attributed the invention of various arts and sciences including medicine (Seiner, 1933). Amongst these are:

- The falcon-headed sun-god Re.

- The wonder-working Isis with her son Horus.

- Ptah, the ancient god of Memphis.

- The ibis-headed moon-god Thoth with ram's horns.

- Sekhmet, the lion-headed goddess.

- Most important was Imhotep, whom we shall discuss separately.

- Seth, a symbol of evil and a cause of sickness and epidemics (he killed his brother Osiris, as mentioned previously).

- Ta-urt, the hippopotamus-shaped goddess, pregnant, capped with the sun disc, the two horns, and the two plumes, leaning on an amulet shaped like Isis's knot, presided over childbirth. She was Seth's wife and at Thebes, where she was worshiped under the name of Apet, she was considered as the mother of Osiris. She was also called the "lady of gods" and the "good nurse". Some of her statues were made hollow and full of milk flowing from her breasts to maintain the lactation of wet nurses. Until now in Egypt, there are many lay people who believe in "evil eyes", which stop their lactation (see Canton).

For more details on Egyptian mythology, see Mercatante, A.S. *Who's Who in Egyptian Mythology*. Clarkson N. Potter, Inc: New York, 1978.

THE EGYPTIAN MEDICAL PERIODS

Dr. James Finlayson (1893) stated in the Faculty of Physicians and Surgeons in Glasgow, on January 12th, 1893:

> Ancient medicine is the title of one of the Hippocratic Books. On hearing its name for the first time, we can scarcely restrain ourselves and smile at such a title being given to a book by one who is termed the father of medicine. When, however, we place the title of his treatise over against the abysmal depths of Egyptian chronology, we feel that Hippocrates belongs to the modern rather than to the ancient period of medicine, just as a pursuit of his works convinces us that he is animated by what we may fairly call the modern spirit.

Some comparison of dates may help us in our estimates of the periods referred to:

Egypt Dynasty I	3100 B.C.
Khufu Dynasty IV	2700 B.C.
Exodus of Israelites	±1350 B.C.
Ebers Papyrus	±1550 B.C.
Berlin medical papyrus	±1400 B.C.
British Museum medical papyrus	±1100 B.C.
Herodotus	Born 484 B.C.
Hippocrates	460-357 B.C.
Septuagint version of the Bible	280 B.C. downwards
Diodorus Siculus	90 B.C. to 30 B.C.
Celsus	Lived in the second century A.D.
Galen	128 or 130–200 A.D.
Clemens of Alexandria	Died in about 220 A.D.

In this text, we will review the cardiovascular medical knowledge in ancient Pharaonic Egypt. We will not discuss Ptolemaic or Coptic medicine, since we are reviewing the hieroglyphic sources only.

IMHOTEP

Imhotep (I-em-hotep) (see Figure 6.1) in the old Egyptian language means "he who cometh in peace" (Canton, 1904) and it is the most appropriate name for a healer of the sick and one who brought solace and courage to many anxious patients. He is called the son of the supreme god Ptah and the Greeks called him Imonthes. He lived during the reign of a famous Egyptian king, Zoser, a Pharaoh of Dynasty III (c. 2980–2900 B.C.). Descended from a distinguished architect named Kanofer and from a mother called Khreduonkh, Imhotep received a liberal education and grew up as an Aristotelian genius. He was distinguished for his vast learning and distinguished achievements. He devoted his life to many impressive pursuits, making him one of the most influential individuals in history.

FIGURE 6.1   Statue of Imhotep (commons.wikimedia.org).

First, he was a grand vizier to King Zoser bearing the heavy responsibilities of the Pharaoh including: chief judge, overseer of the king's records, bearer of the royal seal, chief of all works of the king, supervisor of that which heaven brings, the Earth creates, and the Nile brings, and supervisor of everything in the entire land. Amongst some of the departments of his office were those of Judiciary, Treasury, War (Army and Navy), the Interior, Agriculture, and the general executive.

Second, as an architect, Imhotep, in all probability, is the one who designed for his king the step pyramid of Saqqarah (Saqqarah is probably a corruption of an old Egyptian word *Sakari,* the name of an early god of the dead at Memphis). It is a transition between the *mastaba* tombs and the true pyramid form built later. It is the earliest large stone structure known to history and was destined to become the tomb of King Zoser. Imhotep's name is also associated with the temple of Edfu.

Third, as the chief lector-priest, Imhotep had this important office of Kheri-heh herztep, in whose duty he used to recite from holy texts according to Egyptian faith; he was regarded as a magician, Kheri-heh, he took an important part in offerings made in ceremonies for the dead. By this, he was regarded by the common people as the mediator between the king and the unseen powers of the universe and was supposed to influence the final destinies of the dead.

Fourth, as a sage and scribe, Imhotep enjoyed the reputation of being "one of the greatest Egyptian Sages". His activities were not only in architecture and medicine, but also in general subjects and philosophy and some of them were extant at the dawn of the Christian era. His proverbs were handed from generation to generation and were noted for their grace and poetic diction, so that he was described as the "master of poetry".

Fifth, as an astronomer, Imhotep, like all Egyptians, believed in the influence exerted by the heavenly bodies on the welfare of human beings. His name was reputed to have been associated with the god Thoth (Hermes) in astronomical observations as the study of heavenly bodies, eclipses, precession of the equinoxes, occultation of planets, etc. According to Diodorus Siculus, "there is no country where the positions and movements of the stars have been observed with such accuracy as in Egypt".

Sixth, Imhotep as a magician physician combined the high reputation of being a famous priest-magician of Memphis and also as the most eminent healer of the sick so that in a subsequent apotheosis, he was named demigod and then the god of medicine and compared with Aesclapius of the Greeks (Edelstein et al., 1945). His eminence as a healer of the sick gave him imperishable fame. He was known to have a tender heart toward the suffering humanity. The full apotheosis of Imhotep appears to have taken place in the Persian period (c. 525 B.C.), the year in which Cambyses conquered Egypt. To this new god is now assigned a divine father, Ptah, instead of his human father Kanofer and he became a member of the great triad of Memphis, i.e. Ptah, Sekmet, and Imhotep. With time, he was worshiped throughout the land. In his honor, at least three temples were built. One at Memphis, which became a famous hospital and a School of Medicine and Magic and near this temple, where he had spent his days in relieving pain and prolonging life, Imhotep eventually was laid to rest. The second temple is at Philae and one of the halls was used for clinical purposes thousands of years ago; it was built under Ptolemies. The third temple was built at Thebes. Apart from these three, there are records of a sanatorium, which was situated in the upper terrace of the temple of Hatsehp-sowet at Deir-el-Bahari, dedicated to the worship of Imhotep and to the other healers, Amenophis and Hapi. The Ptolemaic kings built many other small temples, viz. Kasr-el-Agouz, Medinet Habu, Deir-el-Mediheh. Ceremonies were done as festivals every year in these temples to commemorate Imhotep. Pilgrimage to his tomb was frequent and people practiced "incubation sleep" in his adjoining temple and that was followed by many recoveries through therapeutic dreams.

The worship of Imhotep lasted during the Christian era until the second century A.D. in Upper and in Lower Egypt, until probably the fourth century A.D. (30 B.C.–300 A.D.), when due to Roman occupation, the country was impoverished and Christianity spread. The introduction of Greek medicine and culture and the founding of Alexandria in 331 B.C. undermined the ancient medicine and culture. At that time, rapid progress in anatomy, physiology, pathology, and *materia medica* was attained. (For further reading on Imhotep, see Hurry, 1975.)

About Imhotep, Sir William Osler, in 1923, said, "Imhotep was the first figure of a physician to stand out clearly from the mists of antiquity" (Osler, 1923).

King Hadrian favored Imhotep, and in Rome the King had erected a monument in memory of Imhotep. On one side was engraved, "He hears Imhotep, plea of him who calls him he let the need become well". Apart from Imhotep, there is a large list of great physicians and amongst them few female doctor gods. According to Egyptian mythology, the goddess Isis was a healer. The first Egyptian lady doctor was Pesechet and the first in the world (Old Kingdom Dynasty V–VI) (see Nunn, p. 124).

CHAPTER 7

Influence of Ancient Egyptian Medicine on Greek Medicine and Beyond

THERE WAS SIGNIFICANT CONTACT between the Greek and Egyptian civilizations. Not only did Greece enjoy a privileged contact with the land of the pharaohs, touring Egypt later became an obligatory pilgrimage for Greek scholars. Serge Sauneron has rightly emphasized the respect shown by Greeks for Egyptian learning in spite of the reluctance of temple priests to disclose the occult. The rise of rational medicine in Greece may be traced to the year 670 B.C. when the Psammatichus, King of Egypt, opened the land of the Nile to the inhabitants of the Mediterranean and European countries. Before that time, the valley of the Nile was a land of mystery. All that was known about Egypt were vague reports that leaked out through political refugees, who fled to

Egypt to escape punishment or from adventurous Greek pirates. These men stealthily visited the Valley of the Nile and brought back prodigious and miraculous reports of the Great Pyramids, Sphinx, obelisks, temples, etc. The Greeks made Psammatichus attain supreme power over Egypt in the civil war against the Persians. He made Egypt accessible to the Greeks. He opened the Egyptian schools to the young Greeks who were thirsty to learn the "wisdom of the Egyptians". The Greeks imbibed this, made additions to it and diversions from it, and embarked on a new route.

Logan Clendening in his *Lectures* of 1952 and Dr. J.B. de C. Saunders said on this subject (p. 13):

> Indeed many of the early Greek historians ascribe the inventions of medicine to the Egyptians. This tradition would seem to derive ultimately from the verses of the Odyssey calling attention to the great skill of the Egyptian physician and to Helens being taught the uses of the drugs while in Egypt as is remembered in the naming after her of the plant helenion or elecampane. The probabilities of extensive mutual exchanges between Egyptian and Greek physicians are very obvious during the Alexandrian period when the Great Greek schools led by investigators, such as Herophilos (c. 323–283 B.C.) and Erasistratus (ob. c. 304 B.C.) existed in Egypt, but earlier contact with which we are more concerned can be documented as occurring more than two hundred years prior to the Ptolemaic rule in Egypt ... Thales of Miletos (c. 624–544 B.C.), who was a pupil of Egyptian priests, reputably founded the Ionic School of Medicine, and there is undoubted evidence that he was greatly influenced by Egyptian empirical mathematics in his establishment of the beginnings of an abstract geometry. Eudoxos of Cnidos (c. 408–355 B.C.), the great mathematician who developed the theory of the

golden section, is said by Clement of Alexandria to have studied under the Egyptian Chonuphis, which determined not only his great mathematical contributions, but also his thinking as a physician.

The intercourse of Egypt with Syria and the coast of Asia Minor dates from a very early period; every merchant caravan to the many cities of Asia Minor carried the fame of the Egyptian physician. Because of this association Egyptian medical ideas have the greatest influence among the Ionian and Carians, especially at the older school of Cnidos, if not so much at Cos.

After giving many specific examples of transmission, the author concluded:

The philosophical and scientific revolution which occurred in Greece of the sixth century B.C. was not born in a vacuum. Its antecedent was the inheritance not only of the prescientific beliefs, myths, and cosmogonies of the Greeks themselves, but also the accumulated treasures of two thousand years of Babylonian and Egyptian thought ... The contributions of the ancient Egyptian physicians to the continuous stream of thought, which established modern science and medicine constituted a powerful beginning.

(CLENDENING, 1952)

For further information on Greek and Egyptian medicines, see Ghaliongui, Nunn, and Halioua and Ziskind.

Alexander the Great built Alexandria in 332 B.C. Although he died before its completion, it attracted the greatest prosperity under the Ptolemaic rule, which made it the capital of Egypt. The city was as great as its master designer, Alexander the Great.

Apart from its grandeur, it had a library that contained 700,000 volumes in all sciences. It was thus the Mecca for scientists and scholars from all over, especially from Greece, e.g. Euclid, Aristchus, Archimedes, and many others. In medicine, there were two eminent scientists, namely Herophilus (c. 300 B.C.) and Erasistratus (c. 270 B.C.).

Herophilus, often called the father of anatomy, described the brain and its membranes in detail and recognized it as the center of the nervous system as well as the seat of intelligence. He distinguished between the arteries from the veins based on the wall thickness. He named the pulmonary artery, the arterial vein, and the pulmonary veins as the venous arteries. He recognized the power of the heart as the cause of the pulse and he studied its rate and rhythm. He also distinguished the nerve trunk from the tendon and the blood vessels, thus resolving the problem of the *metu*. He studied the anatomy of the liver, pancreas, salivary glands, and the gonads. He described the duodenum and the prostate glands. His book contained several treatises on anatomy and one on the causes of sudden death.

Erasistratus, Herophilus' rival, studied the valves of the heart and described their functions. He incorrectly thought the blood changes into air (*pneuma*) when it reaches the arteries. He studied the nervous system and noticed the difference between motor and sensory nerves and described correctly the atrioventricular valves as the bicuspid and the tricuspid valves and that they are one-way valves. He conceived the heart as a true pump (cf. Bonnabeau, 1983).

Although Herophilus claimed that he was the first to count the pulse by the water clock, J. Breasted noticed that this was mentioned in the Edwin Smith Papyrus c. 1600 B.C. It is believed that the water clock (clepsydra) existed since the time to Thutmose III. Ebers Papyrus in its beginning is entitled "Beginning the Secret of the Physician". This is considered by some authors that counting the pulse was a secret of the Egyptian physician.

Other scholars in the Alexandria School included Callimachus, who classified every field of learning, Philinus of Cos, who studied anatomy and physiology and experimental therapeutics, and Heracleides, who studied human anatomy and developed surgical techniques.

Thus, the Pharaonic and Ptolemaic medicines were in cooperation in the Alexandria School of Sciences (cf. Van Praagh, 1983; Major, 1954; Ghalioungui, 1984). For further reading on Greek medicine in Rome, please see Allbutt (1921). With the fall of the Greek Empire, the Roman Empire rose and ruled Egypt for several centuries.

The Romans inherited medical knowledge from the integration of the two previous empires of the Egyptians and the Greeks. They also developed eminent physicians and researchers such as Galen, Dioscorides, Pliny, Antyllus, Orubasis, Paul of Aegina, and Soranus. With the fall of the Roman Empire, Christianity entered Egypt and the popular language changed to Coptic and was written in Greek characters. For more details, see the chapter on Hippocrates in Egypt.

Along with the downfall of the Roman Empire, Egypt was occupied by the Arabs and the new religion (Islam) in 642 A.D.

By then, Egyptian medicine had gradually changed to quackery relying on deities. The Arabs came from a culture that had no scientific past and they were eager to learn and assimilate the Greek knowledge. The Arab–Muslim scholars kept the torch lit by Egyptian and Greco-Roman scholars burning bright, while Western Europe was still in the Dark Ages. According to Dr. Ralph Majors' book,

> They not only kept alive the learning of antiquity, but they made important additions to it. From the second half of the eighth century A.D. to the end of the eleventh century A.D. Arabic was the scientific, progressive language of mankind. Anyone who wanted to learn up to date had to learn Arabic.

As the Islamic Empire extended from east Asia to Spain, the leaders were enthusiastic about translating the old knowledge. They hired translators from Syria, Persia, and Asia Minor, who translated all Greek medicine to Arabic or individually through Syriac. The most famous of them was Huneyn Ibn Isaac. Jewish and Christian translators made a great effort toward this transition with the encouragement of the Caliphs. Medicine progressed rapidly. Hospitals, clinics, and sanatoria were opened in many places in the empire, especially, Mesopotamia, Egypt, Syria, North Africa, and Spain. New fields of medicine appeared and research in pharmacology was advanced. Arabic/Muslim scholars were from the east of the empire to its west. Amongst these, just to mention a few are Rhazes, Haly Abbas, Abulcasis, Al-Kindi, and Isa-Ibu-Ali. It is important to mention as well Ibn al-Nafis who discovered lesser circulation in the 13th century and his views were received in well-known Arabic treatises; his discovery was recognized by the medical centers, which was well before William Harvey's discovery in the 17th century. Details of his circulation discovery are found in Ghalioungui's text and are summarized by Willerson and Teaff. A copy of his work is found in the Bibliothèque Nationale in Paris and is published in Arabic by Fayyad. Also, not to be forgotten is Avicenna (the Prince of Physicians). He was a savant in medicine, and he compiled about 100 books in different fields; one of his important books is called *The Canon* (*Cannonis Medicis Avicennae*). This book was the text in the medical schools in the Arab world and Europe for many generations. Dr. Osler said it was in great demand at the University of Vienna for 500 years after Avicenna's death (1037). There will be an abstract of his cardiovascular chapters in the next chapters (see Major's book *History of Medicine*).

HIPPOCRATES AND EGYPTIAN MEDICINE

Hippocrates was born in Cos in 458 B.C. When he was 19 years of age, he wanted to go to Egypt in spite of opposition from his parents, so he came rowing with his chest naked and came to Egypt in the hot month (Pyanepsion, which is roughly October). In Egypt, he was well received by the scientific society, which was very influential at that time. He remained three years in Egypt during which he made the acquaintance of the Egyptian scientists.

El-Gammal wrote about Hippocrates' journey to Egypt in his article "The Role of Hippocrates in the Development and Progress of Medical Sciences" (1993). He described the Egyptian civilization that Hippocrates admired, especially the extraordinary development of the Egyptian sciences, particularly the physical sciences. Hippocrates met with physicians who were treating the sick in the temples, where there was the encyclopedia of Thot that included six books about anatomy, pathology, surgical instruments, and diseases of women. He studied not only medical specialties, but also astronomy, which was contained in four books on hieroglyphics, cosmology, topography of Egypt, and he was horoscopy, the latter of which he was particularly interested in. He experimented with animal magnetism, somnambulism, and hypnotism. He visited Memphis and Canope for the temple of Osiris. He remained there for three years, after which he returned to Cos when his grandfather had died.

Back at home, he was the great physician, "the Father of Medicine", the scientist, the artist, and the philosopher. He gathered up the threads of the fabric of ancient medicine and embroidered them into a new scientific field. Admired by all, really understood by few, imitated by many, and equaled by none, Hippocrates was the master of medicine since his time. He ended his eventful life at Larissa most probably in 377 B.C. From his collection, it is evident that he was influenced by Egyptian medicine – both in theory and in practice. There are many of his recipes translated word-for-word from the Egyptian papyri. His collection "On Diseases of Women" and "Sterility" corresponds in an extraordinary way to the verses of the Papyrus Carlsberg and Papyrus Kahun.

SUGGESTED READING

Arab, S.M. Bibliotecha Alexandria: The ancient library of Alexandria and rebuilding of the modern one. *Daheshe Voice.* 2003; 8(4): 7–13.

Ghalioungui, P. Medicine in the days of the pharaohs. *Ciba Symposium.* 1961; 9(5): 206–220.

CHAPTER **8**

The Egyptian Physicians and Their Assistants

I N ANCIENT EGYPTIAN HISTORY, there were no proper medical schools as we have now. The physician had to learn from his father or parents or a relative after finishing his general education. Physicianstreatedpatients until they improved or were completely cured. If the patient died, the physicians performed an autopsy to detect the cause of death. Ziskind and Halioua (2004) documented that physicians were knowledgeable in anatomy by attending the embalming and doing animal dissections and comparing them with human anatomy. For the practitioners, it was important to learn to practice in special institutions, which did not specialize only in medicine, but were comparable to a university where there were many "scribes" for different specialties and where they consulted old literature and documents. Many of the papyri and the *Book of the Dead* came out of these institutions. Amongst the students' programs, there were interchanges of opinions under the direction of older experienced persons. These were called the

"House of Life". One of the famous professors at that time was named Seneh. The invading Persian Cambyses destroyed these schools, but again they were restored by Darius I, who intended to make them useful to the sick and serve as an eternal commemoration of the gods.

According to Herodotus, the physicians were specialized, each one dealing with certain diseases; there were oculists, dentists, specialists of diseases of the head or the abdomen, etc., as mentioned previously. But some of them were general physicians, e.g. Iri, who was the chief of physicians to the Royal Court, did ophthalmology and gastroenterology. Books also were specialized, e.g. the Clemens of Alexandria mentioned a book of six volumes; the fifth was for ophthalmology and the sixth for gynecology.

In Pharaonic times, the doctors had no private work; they were paid by the state. Diodorus of Sicily says: "In the military expeditions and on journeys all people were treated medically free of charge".

The physicians were in many different departments and of different grades, e.g. there were the practitioners, chief physicians, and a chief physician of the South or the North or both. Even in the Royal Court, there were deans of the court physicians, inspector physicians of the court, and chief physicians of the king (cf. F. Jonckheere, 1944).

The Egyptian physicians had to examine their patients very carefully and methodically and treat them ethically (Hastings, 1925). The case was compared with those mentioned in the papyri. If the case was simple, it was directly treated, but if it was complicated, then a document had to be consulted. An example of this was given in the Ebers Papyrus (1875):

Instructions for someone who suffers from the stomach:

1. (Examination): If you examine a patient having some trouble of his stomach in which he feels it very difficult to eat anything, having the abdomen tight and the heart accelerated,

unable to walk like a man suffering from inflammation of his arms, then you have to examine him lying down.

2. (Diagnosis): If you find his body hot, with discomfort on the stomach (here meant the epigastrium), then you will say this is the result of his liver.

3. (Treatment): Then you resort to secret medicines of the plants which are used in a similar case: Plant *pakhseret* and date stones, you mix and filter with water and the patient has to drink in the morning for four successive days, then you palpate again his abdomen. Then a second and third visit are not paid until the patient is cured or discharged. Whether the physicians prepared the prescriptions to their patient is not clear. But it was found that the country physicians were also the pharmacists who used to prepare the medicines for their patients.

The paramedical corps were the priests of Sekhmet (lion-headed goddess) and some of them were combined physician—priest; many of them have been mentioned in the literature. There were also veterinary physicians amongst them, and some were specialized also as, e.g. "Ahanekht", who was a specialist in treating bulls and among them, there were also priests (*ouabou*).

In addition, there were nursing men, masseurs, physical therapists, pharmacists, and bandagists called *out*. Sais mentioned that there was also a school, which prepared the girls to be midwives as mentioned previously. There were two midwives assisting Hebrew women in Egypt mentioned by name in the bible (Exodus 1:15). There were famous physicians: Stemetnach before Imhotep, Sesa was a famous surgeon, Niankh – Re in Dynasty IV, Pepi – Ankh in Dynasty VI, and Iwti in Dynasty XIX. There were female physicians; for more details, please see Ghalioungui's book *The Physicians of Pharaonic Egypt* (p. 92). There were at least 100,000 known physicians. Nunn tabulated 150 dignitaries or well-known physicians and their dynasties. According to Stetter, for each 1 (physician) recorded, there are 1,000 forgotten.

SURGICAL PRACTICE IN ANCIENT EGYPT

Egyptian surgical practices were elegantly summarized by Ghalioungui (1963), from whom we present a brief abstract.

There were surgical specialists in ancient Egypt similar to other specialties. Surgical descriptions are noted especially in the Edwin Smith and Ebers papyri. They treated clean wounds by suturing and by adhesive linen tape soaked with gum Arabic (comparable to the steri-strips of today). Infected wounds were cleaned and molded bread and different poultices made a dressing. When a wound was oozing blood, they used fresh meat to stimulate coagulation (compared with muscle graft used years ago to stop the bleeding). For ulcers, they used alum, zinc oxide, copper sulfate, many herbal and animal poultices, or extracts. For head injuries, they used trephine, but the method was not clear; probably they used a chisel and hammer. I had the pleasure of seeing two trephine skulls in the Anatomy Museum of the Cairo Medical School by Professor A. Batrawi (1929–1934), who did numerous skeletal archaeological studies. One of them, according to Ghalioungui, was the skull of Horsiesnest Meritaten. Regarding spinal injuries, the Edwin Smith Papyrus was the first document of spinal cord injuries and a typical picture of quadriplegia (Eltorai, 2003). Extremity fractures were described and adequate splints were used until the fracture healed. For minor injuries, they used bandages, poultices, and herbal and animal products. Circumcision was a frequent surgery done by priest circumcisers; they shared this operation with the Israeli rabbis. Abscesses were drained by incision; tumors were removed by using surgical instrument, e.g. knife, scissors, pincers, and other instruments, which were made by special technicians. Hernias were described and were treated by reduction and sometimes by surgical heat cauterization. Professor J. Bitschai described urological surgery and instrumentation in his book *A History of Urology in Egypt* (1956). See also Ghonem for urology. Gynecology and obstetrics are best recorded in the Alkahun Papyrus.

SPECIALTIES

In ancient Egypt, there were generalists and specialists; the former were at a lower level of education than the specialist. They treated patients for general ailments, such as headache, cold, cough, indigestion, diarrhea, tenesmus, nausea, and so forth. Internists specialized in different fields such as gastroenterology, cardiovascular, rheumatology, sleep disorders, and psychiatry. Other independent specialties were ophthalmology; dermatology; dentistry; urology; breast specialists; gynecology; obstetrics; ear, nose, and throat specialist; pediatrics; and geriatrics.

CHAPTER 9

The Sources of the History of Ancient Egyptian Medicine

The Papyri

T HE ANCIENT EGYPTIANS APPARENTLY began to record their
knowledge as early as the third millennium B.C. They developed hieroglyphic writing, which was used at first by carving on
stone. Later, when papyrus was developed, with a brush-and-ink
writing, a cursive script evolved, which is known as hieratic. This
was written from right to left, with red ink for the headings and
black ink for the actual text. Papyrus rolls were made by inter-
weaving split river-reeds (*Cyperus papyrus*), pounding the crude
mats under water, and drying them to form coarse brownish
sheets. After being written upon, they were brushed with a deli-
cate brush made from a frayed reed and were glued together at
the edges to make a roll. These papyri form our chief source of
Egyptian medicine history. Apart from these, some historians

such as Herodotus (450 B.C.), Manetho (300 B.C.), Diodorus (60–57 B.C.), and Strabo (27 A.D.) provide more information. In addition, the monuments also provided a guide to medical study. For further reading, please see *Catalogue of Egyptian Religious Papyri* (1938, p. 82) and Grapow's *Altaegyptischen Medizinichen Papyri* (1935a, p. 83).*

The Egyptians used to write also on linen, leather, and parchment, and on pottery (called Ostraca). The papyri were of different types: royal – the best quality for kings and priests as well as ministers, hieratic – for books, and demotic – for different acts in civil or military life. Before writing, they were dipped in cedar oil to preserve them from decay and from worms. They were then carefully preserved in wood cylinders hardened by heat to prevent humidity and then kept in jars of burned earthenware.

THE MEDICAL PAPYRI

Ebers Papyrus

This was discovered by some peasants in a Thebian tomb in 1862 and was acquired by Georg Ebers (1837–1898) at Leipzig University (see Figure 9.1) (Ebers and Stern, 1875) and reproduced by him in 1875. H. Joachim (1890) in facsimile form made a preliminary attempt at translation in German in 1890. Walter Wreszinski (1913) made an extended analysis in 1913, B. Ebbell (1937) published an English translation in 1937, and an analysis and comment were published by Ghaliongui in 1987. Its length is 20.23 meters and its breadth 30 cm, and it is composed of 108 columns, each of 2–22 lines with few exceptions; 2,289 lines in all. In its numeration of columns No. 28 and 29 were missing without interruption of the text and it ended with column 110, which was a lucky number to the ancient Egyptians that they considered as the ultimate age of human longevity. On the back, there is a calendar

---

* *The most comprehensive study of the Egyptian papyri is within the Grundriss der Medizin der alter Ägypterin produced in Berlin University between 1954–1973 by a team of scientists under the leadership of Professor Grapow in conjunction with von Deines and Westendorf published by Akademie-Verlag, Berlin. It contains nine volumes.*

FIGURE 9.1   Ebers Papyrus (commons.wikimedia.org).

assuring its date, viz. in the 16th century B.C. (Dynasty XVIII, during the reign of Amenophis III). But it is not original since it is evidently compiled from one or more other books many centuries older. Indeed, the document itself states that some portions date from Dynasty I and, therefore, was in part at least in existence during Imhotep's lifetime. It is kept in the University of Leipzig Museum. It is written in hieratic language. It contains a long list of prescriptions for numerous ailments and specifies the remedies to be used as well as the doses and the mode of administration. There is also considerable progress in clinical examination, diagnosis, and therapeutics. There is also an accurate knowledge of the skeleton and the methods of the successful treatment of fractures. The positions of the stomach and intestines were known. In the Ebers Papyrus, a lengthy description of the vascular system and the widespread action of the heart, the circulation of blood, and anatomy of vessels are given and will be discussed in detail. The following is a classification of this important papyrus in columns:

1–2	Recitals
2–55	Compilation for remedies for internal diseases
55–64	Prescriptions for the eyes
64–76	Recipes for the skin
76–85	Recipes for diseases of the arms and legs
85–93	Formularies for various conditions
93–98	Formularies for female conditions and household matters
99–103	Two monographs on the heart and blood vessels with glosses (cardiovascular text)
103–110	Treatment of surgical conditions

There are 829 prescriptions with little duplication using 320 different ingredients. According to Leake (1952), the Ebers Papyrus deserves much more extensive study and annotation than it has so far received. Much still remains to be accomplished in the identification of disease conditions and drugs to which reference is made. Amongst the diseases that have been identified are fevers, dysentery, intestinal worms, hematuria, dropsy, heart disease, rheumatism, liver disease, polyuria (diabetes?), intestinal obstruction, a large number of gastrointestinal troubles, cough (abscess of lungs?), burns, ulcers, infected nails, wounds, impotence, scurvy, E.N.T. diseases, dentistry and stomatology, breast diseases, skin diseases, gynecological troubles, erysipelas, tumors, bubonic abscesses, vascular diseases, hernias, and angina pectoris.

EDWIN SMITH PAPYRUS

This was written around the beginning of Dynasty XVIII (c. 1600 B.C.). It was acquired by Edwin Smith, the first

American Egyptologist, who was a farmer in Thebes interested in excavations; he learned Egyptology in London, Paris, and Cairo. In 1862 and on his death, it was presented by his daughter to the New York Historical Society in 1906 and was later transferred to the New York Academy of Medicine. In 1930, Professor James H. Breasted made a detailed study in two volumes with comments with the help of his colleague Dr. A.B. Luckhardt, Professor of Physiology at Chicago University. It comprises 17 columns of hieratic writings with four-and-a-half columns on the verso. A skillfully organized surgical text, it contains 48 typical cases. The work was left unfinished by the scribe in the middle of the 48th case, which deals with injury of the spine. The cases are presented in a purely scientific manner and do not involve any kind of magic. The material on each case has a heading written in red ink, then the examination followed by diagnosis, prognosis, and treatment. It deals with the heart and the counting of the pulse by using water counter or comparing with the examiner's pulse. Breasted emphasized that this papyrus was copied from a much older document as early as 3000 B.C. It also contains typical anatomical studies, especially in relation to the brain and meninges. Nervous symptoms as evidences of fractures of the skull or spine were well demonstrated. It is the first document in history describing spinal cord injury (Eltorai, 2010). Manipulative reductions are described in the best useful way. The second part contains eight incantations to keep aside for the wind of epidemics, and the third part is dedicated to rejuvenation. The Smith Papyrus presents convincing evidence that cyanosis was observed and ancient Egyptians described it. (Sign of congenital heart disease?) The earliest illustration of a blue-skinned person was taken from the tomb of Osiris, dating to about 2000 B.C. For more details on the Smith Papyrus, see Ghaliongui's book *Magic and Medical Science in Ancient Egypt* (1936, pp. 58–66) and Hussein's book *The Edwin Smith Papyrus*.

The papyrus contains 48 surgical cases mostly traumatic. However, there is a portion which is cardiovascular. According to Breasted:

> the ancient surgeons knew of a cardiac system and it was surprisingly close to recognition of the circulation of the blood, for he was already aware that the heart is the center and the pumping force of a system of distributing vessels.

He was already conscious of the pulse (rate and strength). The pulse was counted late by Herophilus in Alexandria of Egypt by a water clock. The concept of the failing heart was also documented (see Breasted). Professor Myerhofer wrote: "The author of the book of surgery wrote his work in a tremendously methodical manner" (quoted by Stetter).

HEARST PAPYRUS

This was discovered by the American Hearst Mission of California in 1899 at Deir El-Ballas and published by George Reisner in 1905 (Reisner, 1905). It dates back to Dynasty XVIII, about the time of Thutmose III (1540–1450 B.C.), probably a little later than the Ebers Papyrus. It is written particularly for a general physician. It contains a bigger number of invocations and recitals than the Ebers Papyrus. There are 250 prescriptions for a variety of conditions and many of them are found in the Ebers Papyrus and Berlin Medical Papyri, e.g. skin diseases, heart, chest, bladder, etc 17 cm high × 3.5 meters long.

THE KAHUN PAPYRI

This is probably the most ancient; it dates back to Dynasty XII (c. 1950 B.C.) at the time of Amenemhat III in the 29th year of his reign during the Late-Middle Kingdom. It was discovered in winter (1888–1889) at El-Lahun (near Fayoum), a town that flourished. It was found by Flinders Petrie and was erroneously called Kahun. It is so mutilated and fragmented that its interpretation was most difficult, and it is housed at University College of London. The English translation includes passages into Latin by Griffin in 1898 and another translation by Stevens in 1975. It is composed of 34 paragraphs describing women's diseases, not necessarily of the reproductive organs, and diseases of the teeth and gums.

The medical part contains 17 gynecological prescriptions taken by mouth, pervaginum, pessaries, or fumigation. The third page contains 17 prescriptions concerning the assessment of sterility and pregnancy and the ascertaining of the sex of the unborn child. In paragraphs 21 and 22, there are pessaries to prevent conception using the crocodile excrement, honey, and sour milk or sour milk alone, or gum acacia. The second section is veterinary, and although it is the oldest document in this branch of medicine, it is a copy from an older one. It is carefully written and contains prescriptions. The third section is mathematical.

THE BERLIN PAPYRI

In the Berlin museum, there are two papyri, the smaller one is a fragment which dates to 1450 B.C.; it contains prescriptions and magical recitations for the protection of mothers and children and also used for the treatment of the latter. So it is the oldest known document on pediatrics. It covers 15 columns. The larger fragment was found by Giuseppe Passalacqua, a curator in Egypt, near Saqqara, and was sold in 1827 as a part of a collection of antiquities to Fredrich Wilhelm IV of Prussia for the Berlin Museum under #3038. The papyrus was found in an earthenware vessel. It dates to about 1350 B.C. during the reign of Ramses II (Dynasty XIX). It was first translated by Burgsch into German and carried his name. A full translation was done by Wreszinski in 1969 in German. It contains 24 columns, 21 recto 1–191 paragraphs (on the front), and three verso 192–204 paragraphs (on the back). It is not systematically arranged and there are many errors in copying. It contains 170 prescriptions for worms, diseases of the breast, heart, diseases of legs, vessels, abdomen, ears, hematuria, and rheumatism. It also describes fertility, the diagnosis of pregnancy, and the possible sex of the unborn child. Many of the prescriptions are the same or similar to those written in the Ebers or Hearst papyri (cf. Wreszinski's *Der Grosse Medizinache Papyrus*, 1909). The Berlin Papyrus 1 is small and kept in the Berlin Museum under #3027. It is a small fragment dating to Dynasty XVIII. It contains magico-religious medicine and is on the care of mother and child after birth. Erman translated the papyrus in 1901.*

THE LONDON MEDICAL PAPYRUS

It is kept in the British Museum under #10059 and was written at the end of the XVIII Dynasty around 1350 B.C. close to the reign of King Tut. Walter Wreszinski annotated it in 1912 (Wreszinski, 1912) in German, and it is included in the Grundriss. It is not preserved in good condition and was copied from older ones. It contains 29 columns, 19 recto and 10 verso. In addition to many recitals, it contains 61 prescriptions and concludes with prayers. There is a small section on gynecology and 17 paragraphs parallel with Eber's.

CHESTER BEATTY PAPYRI

The well-known English Egyptologist A.H. Gardiner transliterated this papyrus in 1935. It was translated and annotated by F. Jonckheere in 1947 (Jonckheere, 1947). It was discovered in Deir el-Medina in 1928. It appears to have been written in the Dynasty XIX about 1200 B.C. It consists of eight columns, the beginning and end of which are damaged, and is part of a treatise on recipes for diseases of the anus. It is the oldest document on proctology, as it seems that there were specialists in proctology in ancient Egypt. Most of the recipes are not found in other papyri. It also contains one case description. It was given by Sir Alfred Chester Beatty to the British Museum and bears the papyrus #10686. The Chester

---

* Erman A. Zauberspruche für Mütter und Kind (from Papyrus 3027 in the Berlin Museums), Berlín, 1901, microfiche.

Beatty Papyri #s 10, 15, and 18 (British Museum Papyri #s 10690, 10695, and 10698) contain medico-magical prescriptions and spells, one being wholly concerned with aphrodisiacs. The #6 recto is concerned with scorpion stings and #15 has two prescriptions for destroying thirst in the mouth.

CARLSBERG PAPYRUS #8: 13 FRAGMENTS

Their origin is not known; they date to Dynasty XIX or XX and are kept in the Institute of Egyptology in Copenhagen. It was translated and annotated by E. Iversen in 1939 (Iversen, 1939), and later in the *Grundriss*. The Carlsberg Papyrus is mostly about the detection of pregnancy, the sex of the unborn child, and the ability to conceive.

THE RAMESSEUM III, IV, AND V PAPYRI

In 1896, Quibell found 17 papyri in a wooden box at the bottom under the brick magazines behind the temple of Ramesseum. These were probably written at the time of the Kahun Papyrus, i.e. 1900 B.C. There are references to Amenembrut III, whose reign was about 1854 B.C., in papyrus six. The hieratic text was published by Gardener in 1955 and hieroglyphic by Burns in 1956 and to the *Grundriss*.

Ramesseum Papyrus III contains 31 paragraphs in section A and 34 in section B. The medical part discusses diseases of the eyes, gynecology, and diseases of children.

Ramesseum Papyrus IV contains similar prescriptions as those of Kahun concerning labor, the protection of the newly born, and the judgment of its vitality. There is one contraceptive formula made of crocodile dung.

Ramesseum Papyrus V had its beginning and end destroyed and it contains 20 prescriptions mostly dealing with the relaxing of stiff limbs, many of which are found in the

Ebers and Hearst papyri. It is written in hieroglyphic script, not in hieratic. This arrangement recalls the writings on sarcophagi and probably is a witness to the archaic origins of these two papyri, the oldest medical ones in our possession.

THE BROOKLYN PAPYRUS

The Brooklyn Papyrus is housed in the Brooklyn Museum in New York City. It is dated probably back to the 30th dynasty or perhaps to the early Pharaonic era. It is written in idioms of early classical Egyptian language. It was translated into French by Sauneron in 1989. It is written exclusively for snake and scorpion bites. It has an upper and in the lower part, it contains medicines and spells for the treatment of the snake and scorpion bites.

CROCODILOPOLIS PAPYRUS

The Crocodilopolis Papyrus was written in the second half of the second century A.D. and is in demotic script; it is entirely free of magic and incantations of any sort. It lists remedies for a variety of ailments. It combines classical Egyptian drugs alongside new-found remedies from the Mediterranean area, which were not mentioned in previous Pharaonic medicinal papyri. Crocodilopolis was a city southwest of Memphis along the Nile. The papyrus is in Vienna.

THE LEIDEN PAPYRUS

According to Estes, The Leiden Papyrus is a collection of spells for a wide range of purposes (not all of them for healing). Written in the third century A.D., some of them are dated from the New Kingdom. There are incantations for many strange conditions. Spells are for gouty foot and dog bites, which they treated by cleaning the wound and dressing it with honey. Some spells are written to cause disease, for example, to make a man blind, while others are aphrodisiacs. Some medicines were translated as opium, hyoscyamus, madragora, etc.

The A.D. papyri although are not Pharaonic are mentioned for completion of the subject on medical papyri. The museums of Paris, London, Turin, Berlin, Budapest, Leyden, the Vatican, Cairo, and elsewhere have considerable numbers of magical papyri, which although not generally therapeutic in character are medical in so far as their objective is the treatment and cure of disease, personal injury, and attacks of venomous animals.

Ostraca

Apart from the papyri, prescriptions were found or isolated prescriptions are rare, but few are written on limestone flakes or potsherd, e.g. the Louvre Ostracon (3255) contains two prescriptions for the ear; the Ramesseum Ostracon 35 for protection against snake bite; the Medineh Ostracon 1091 is a prescription for a "cure-all" in any part of the body; and also, the Milne Ostracon for the protection of the limbs on the principle of "Like influences like", e.g. homeopathy.

STELAE

In the later period of ancient Egypt, stelae were found in shrines, temples, and homes; they were known as *cippi*, which were used to convey protection from attack by certain animals, for example, snakes, scorpions, and crocodiles. One of them shows the young Horus victorious over a crocodile. The most complete and well-known *cippi* is found in the Metropolitan Museum of Arts in New York. For more details, see Nunn pp. 107–109. Other protective incantations and amulets are found in theBook of the Dead.

Other sources include such documents as the Papyri of Coptic and Greek material, which represent the Pharaonic Medicine and are dispersed in the World Museums.

The papyri can be tabulated according to their dates as follows:

Name	Date	Condition	Contents	Location
Kahun	1900 B.C.	Fragmentary three sheets	Unrecognized on women's diseases and pregnancy	University College, London
Edwin Smith	1600 B.C.	Unfinished 17 columns, 4 on verso	Well-organized surgical text 48 typical cases. Cosmetic recipes on verso.	New York
Ebers	1534 B.C.	Complete in 108 columns (numbered 110)	Medical text, hundreds of recipes classified by disease, few cases on anatomy monograph	Leipzig
Hearst	1550 B.C.	Incomplete, 18 columns	Poorly organized practitioner's recipe book	California
Berlin ii	1550 B.C.	Nine columns, six on verso	Popular charms for childbirth and care of infants, two prescriptions	Berlin Museum #3027
London	1350 B.C.	Fragmentary, 19 columns	Recipe book with recitals	British Museum #10059
Berlin i	1350 B.C.	21 Columns, 3 on verso	Recipes, recitals, signs of pregnancy	Berlin Museum #3028
Carlesberg viii	1300 B.C.	Recto and verso; recto is badly damaged	Recto is mainly concerned with diseases of the eyes, verso is gynecological,	Copenhagen
Chester Beatty vi	1200 B.C.	Incomplete, eight columns	Anal diseases, one case report	British Museum #10686
Ramesseum iii, iv, v	1700 B.C.	The medical portions of the papyri were badly damaged	Gynecological, medical, magical	Oxford
Brooklyn	300 B.C.	Beginning and end missing, middle is in generally good condition	Snakes	Brooklyn
Crocodilopolis	A.D. 150		Remedies	Vienna

CHAPTER 10

Ancient Egyptian Medical Sciences

ANATOMY

The first knowledge of anatomy was described by ancient Egyptians although it was not in the terms of anatomical dissection as it is currently studied and which was later developed in the Greek era. Manethaton, as mentioned by Nazmi (1903) and by Canton (1904) report that Athothis, the son and successor of Menes, the founder of Dynasty I about 3400 B.C., was a physician and wrote six books on medicine, the first of them being written on anatomy and that Tsartros, the second Pharaoh of the Dynasty III, was interested in medicine and anatomy.

That the Egyptians possessed at least a basic knowledge of the structure of the human body may be assumed from the technique of the Egyptian embalmer.

The special treatment of the viscera acquainted the Egyptian physicians with the appearance and position of the organs and enabled them to recognize the homologies between the internal organs of the human body and those of animals; the latter, having been long familiar to them through the practice of cutting up animals for sacrifice and food. It is a noteworthy fact that the various hieroglyphic signs represent pictures of the organs of animals and not of human beings. This shows that they had some knowledge of comparative anatomy, recognizing the essential identity of structure between man and higher animals. In some instances, they borrowed the signs based upon the organs of animals and used them unaltered when speaking of the corresponding organs of the human body. Thus the sign for the heart showed a heart of an ox and the sign for human throat was represented by the gullet of an animal. The sign "womb" is a configuration of the bicornuate uterus of the cow, for "ear", a nonhuman mammalian ear, and for the tooth, a mammalian tusk. Other hieroglyphic signs borrowed from mammalian anatomy include those for the liver and the mammae of the hoofed animal (cf. Champolion, 1824). (For further reading on hieroglyphic signs and syllables, see Posner, 1959.)

In the ancient Egyptian language, there are over 100 anatomical terms. On the whole, the gross anatomy of the body is fairly accurate. However, there is failure to differentiate between nerves, muscles, arteries, and veins. One word, *metu*, denotes these structures, and the Egyptians appear to have regarded them as a single system. The same term was applied to the vessels communicating with the heart. This appears to be a generic term indicating any tubular structure whether solid or hollow and whether it contains blood or otherwise. Until now, the Egyptian public gives the word *irk* in Arabic to indicate any of these structures. This sometimes causes confusion when one takes a case history or receives the patient's complaint by correspondence. The same word is applied, for example, to the temporal vessels, the sterno-mastoid muscle, tendons of the hand or leg, a varicose vein, an aneurysm, even a hernia going down into the scrotum. So, it seems this word was used popularly and was kept in the physician's description and was passed down through the ages to the present time. For more on the anatomy of the heart, see Grapow's *Über die Anatomischer Kenntrisse den Alten Ägepteschen Arzte* (1935).

A lot of anatomical information is obtained from the Edwin Smith Papyrus; there is a description of the skull sutures and fontanelles. They described the meninges and the underlying brain. The anatomy of the nasal bones, the zygoma, the temporal bones, the mandible, and the temporal mandibular joints were described. The vertebrae and the complete spinal column and the long bones of the extremities were described.

PHYSIOLOGY

General Concepts: the Role of Spiritual Powers in Human Physiology

The Egyptian conception of the dynamics of the human physiology is complicated and confusing because of the belief that it was governed by a number of deities and it was assumed that there are other spiritual forces controlling it.

The *Book of the Dead* divides the human body into 36 parts, the functions of each part or organ being controlled by a special deity. Hathor, for example, governs the eyes, the ears is by Asud, the lips by Anubis, the growth of the hair by Nei, and the face by Ra. Thoth, who supervised the functions of the entire body, coordinated all these individual powers. This last divinity was the center of coordination, which we now locate in the brain. There was the *ka*, which governed perception and pleasure and was supposed to go with the body into the tomb and remain there as long as the body was intact or mummified. The *ka* retained consciousness after death. This spiritual entity was represented by the shadow that follows a human being and was supposed to be born at the same time as the newborn infant. On the sarcophagus of Panahemisis is inscribed: "The Ka is the God, he parted not from thee and saw thy shouldst live eternally". Here, the *ka* has become a divine principle, parallel with the Doctrine of the Logos; in its early development, the *ka* was believed to go to Osiris to join the company of the gods (cf. Aristotle, 1959).

Another spiritual phenomenon was *ba* or *bai*, which was symbolized by a bird with a human head. It was the nearest approach to the Greek conception of the disembodied soul. The *ba* was assumed to go to "Hades" after death. For a time, it would wander in the cemetery requiring food. It was associated with the mummy (*sabu*) as the *ka* was with the body in the grave (*khat*). In other words, it represented the personality after death. The third spiritual entity was the *abi*. She controlled the will and the intuition, which were thought to come from the heart. *Abi* was frequently used as a metaphor for the heart. The physical heart was known as *hati*. So, *abi* or *ib* was used for a spiritual meaning, whereas *hati* was used for the organ itself, which the Egyptians knew to be in the left side of the chest and it showed itself by "speaking" (i.e. pulse). *Abi* was used for the glosses. But in some of the texts, the words *hati* and *abi* or *ib* are interchangeable. According to Ebbell (1937), *ib* also indicates the cardia of the stomach, but Lefebvre (1956) thinks that the stomach is (*ro-ih*) or (*ro-in-ih*), which literally means "the opening of the cardia". It meant not only the cardia, but also the fundus of the stomach. In one of the writings of Dynasty XXII, it is stated that "The heart '*ib*'" is a god, whose chapel is the stomach (*ro-ih*), and this latter will be glad when the other members (*haou*) are feasting". A scarab of the cardiac shape was put inside the mummy to replace the heart or the *ab*.

All human physiology in ancient Egypt depended mostly on natural powers. They considered that man's physique is a miniature world (microcosm) and composed of four elements: body solids (Earth), body fluids (Nile), heat (Sun), and breath (Wind). This conception was held and modified by the Greeks later.

PHYSIOLOGY OF RESPIRATION

The entrance of the air into the body through the respiratory channels was well described. The vessels were the channels to carry air (*pneuma*) to the whole body. In the Ebers Papyrus, as we shall see later, a distinction was made between good and bad air and the air of life and air of death. This possibly denotes inspired air (life) and expired or vicious air (death), i.e. CO_2 is retained.

PHYSIOLOGY OF DIGESTION

The stomach and the digestive system prepared the food before it was delivered to the blood by the heart. The most important cause for illness was thought to be overeating. Other causes were either spirits or worms (real or imaginary) in the system. (The parasitic theory is a scientific one as of today's, while the spirits, unseen, must have been the bacteria of the modern times.) Stoppage of the intestinal system seemed to be particularly dangerous; hence, the care of the bowels received great attention and the mysterious specialty of the *Guardian of the Royal Bowel Movement.*

PHYSIOLOGY OF REPRODUCTION

The Egyptians had a curious knowledge concerning the uterus. The organs of reproduction were believed to be a separate entity independent of the other organs and subject to wandering about the pelvic cavity (pelvic nerves). They discussed, therefore, different malpositions. They described hemorrhage, menstrual disorders, tumors, and inflammations. They used local applications and fumigations. They were interested in the diagnosis and management of pregnancy. For contraception, they prescribed: "Acacia spices to be ground fine and with dates and honey rubbed on a wad of fibers and inserted deep into her vagina". Thousands of years later, it was found that acacia spikes contain a gum, which contains lactic acid when dissolved in a fluid. Nowadays, many of the contraceptive jellies contain lactic acid. Diagnosis of pregnancy as in Papyrus Carlsberg VIII and Kahun:

> Also (recipe) to distinguish a woman who will conceive from one who will not: You shall let a clove of garlic ... remain the whole night (in her womb) until dawn. If the smell is present in her mouth, then she will conceive, if (... not) she will not conceive.

Reading from Iversen's translation (1939), to know the sex of the fetus: "You must put wheat and barley in cloth bags, the woman is to urinate on each daily ... if both germinate she will bear, if the wheat germinates she will bear a boy. If neither germinates she will not bear". In 1933, Julius Manger at the Pharmacological Institute in Würzburg, Germany demonstrated that the urine of pregnant women who later give birth to boys had actually accelerated the growth of wheat, while that of women who later bore girls accelerated the growth of barley, but the results were not consistent.

Labor was managed on a birthing stool (Manascha I, 1927). Diseases of pregnancy are described in the Kahun Papyrus, e.g. bladder disturbances, troubles of the legs (? phlebitis), and even cancer.

These data point out that the Egyptians had an empirical knowledge about the circulation by means of which the garlic odor is carried from the uterus to the lungs, especially when the vascularity is increased by pregnancy. The Egyptians thought that pregnancy resulted from the male and not the ovum. According to Ghalioungui (1963), "what happened to the seed could not have been known to them, therefore the common expression 'he was still in the egg' or "he came out of the egg" could not have referred to the ovum".

PHYSIOLOGY OF THE NERVOUS SYSTEM

Although the ancient Egyptians knew the brain through head injuries or through embalming, they did not recognize its true functions, which were attributed to the heart (ib). The Edwin Smith Papyrus gives valuable insight into the role of the spinal cord, in the transmission of information from the brain to the lower parts of the body, typically in #31; the peripheral nerves were documented under *metu*.

The physiology of cardiovascular diseases will be discussed separately.

PATHOLOGY

Pathology, in its current form, was not known to the ancient Egyptians, although clinically they diagnosed illnesses in different organs and different parts of the body.

The areas of pathological knowledge of the ancient Egyptians were:

1. Trauma: Injuries from accidents, labor, and warfare were very well known to them. They described and treated fractures and dislocation of the limbs. They described different types of head injuries including depressed fractures and brain injuries. Based on the Edwin Smith Papyrus, the Egyptians classified the injuries systematically. Injuries of the vault, the occiput, the frontal, the temporal region, the nasal bones, the zygomes, and the mandible were described. As mentioned previously, trephine surgery was performed. Brain injuries were also mentioned and described as well as its membranes and the cerebral spinal fluid, which leaked in open injuries. Glosses were given to a different structure in hieroglyphic (see Nunn, 1996).

2. Spine: Injuries of the spine and spinal cord were described, especially those with the features of quadriplegia.

3. Fractures of the extremities, soft tissue injuries, and internal injury were described.

4. Burns, snake bites, and scorpion and other animal injuries were described.

5. Pathological lesions of the internal organs, namely the heart, the stomach, the liver, the spleen, the bowels, the rectum, and anus were clinically described and treated.

6. Diseases attributed to gods constituted a big section and were clinically treated.

7. Parasitic diseases (see paleopathology) were described.

8. Infectious diseases, especially, tuberculosis (see paleopathology) were described.

9. Neoplasia (see paleo-oncology) were described.

The internal organs were known to the Egyptian physicians, their sites, and, to some extent, their functions and their illnesses as described in the papyri (see in the papyri chapter).

SYSTEM DISEASES

Musculoskeletal system diseases included osteoarthritis, deformities, and other degenerative diseases.

Nervous system diseases included mostly traumatic injuries to the brain and spinal cord, possible facial palsy, poliomyelitis, and leprosy of the peripheral nerves.

Gynecologic system diseases included menstrual disturbances, uterine bleeding, abortion, and possible cancer, uterine malpositions, and sterility.

PALEOPATHOLOGY

Diseases in ancient Egypt have been discovered by paleopathology in the past 200 years. In modern times, new techniques are used, especially, in the mummy projects; the most-known one is the Manchester project in the United Kingdom. The body used to be unwrapped, widely dissected, and mutilated. Nowadays, only small samples are needed and can be used for various studies. The tissues are examined by light and electron microscopy, microbiology, immunology, and DNA patterns. Radiological exams are done by plain radiography, CT scans are commonly used; however, MRI is not possible due to extreme desiccation. Endoscopy is also used with fiber optic equipment.

The diseases of Pharaonic Egypt can be classified into:

1. Degenerative diseases:

    a. arteriosclerosis (fully described later).

    b. osteoarthritis of the limbs and the spine.

2. Parasitic diseases:

    The medical papyri are rich in pathology and antiparasitic therapy (see Nunn).

a. Filariasis: the disease is transmitted by mosquito bites and the adult worm grows in the lymphatics, especially, in the pelvic area leading to obstruction, lymphedema, tissue hyperplasia, and fibrosis (elephantiasis).

b. Guinea-worm (*Dracunculus medinensis*) is a historic infestation in ancient Egypt. A calcified male worm was found in a mummy in the Manchester program. It was also described in the Ebers Papyrus (#875) including its delivery out by winding it on a wooden rod.

c. Schistosomiasis was endemic with both urinary and intestinal forms.

d. Strongyloidiasis was found in the wall of the intestine in the Manchester program, but was not described in the papyri.

e. *Ascaris lumbricoides* (roundworms): Cockburn et al. found evidence in an unwrapped mummy in the United States, in 1975.

f. Tapeworms (*Taenia saginata, Taenia solium*): Hart et al. found the ova in a mummy in Toronto, 1977.

g. Faciola hepatica and Trachinella spiralis (see Estes).

3. Bacterial diseases:

a. Nonspecific bacterial infections such as abscesses, boils, and infections of the internal organs are under investigation by the use of new techniques for antigenic modulations. However, there are established specific infections.

b. Specific infections:

1. *Clostridium tetani* has not been proven, but a possible case of lock-jaw was described in the Edwin Smith Papyrus.

2. Leprosy (Hansen's mycobacterium) has not been described, but few cases of what is called Khonsu's

tumor (nodular form of the leonine face) may affect peripheral nerves leading to paralysis, atrophy, and ulceration. Two cases are suggestive in the Eber's Papyrus (#874 and #877).

3. Pasteurellosis, producing bubonic plague, has not been found in mummies and not mentioned in the papyri.

4. Syphilis did not exist in Egypt from the predynastic era to the Greco-Roman era.

5. Tuberculosis in ancient Egypt is widely studied. Zink et al., in 2003, reported on molecular studies of human tuberculosis (TB) in different periods of ancient Egypt. Molecular identification of TB was from the vertebral bone tissue from mummies and from the predynastic, early dynastic, Middle Kingdom, New Kingdom, and late periods. That encompassed the period from 2100 B.C. to 500 B.C. in nine cases of macromorphological tuberculous (mTB) spondylitis; six out of nine cases were positive for microbacterial aDNA. Out of the 24 cases of nonspecific pathology of the spine, five had positive aDNA, i.e. 20.8%, out of 15 normal cases, seven were positive, i.e. 40%. There was a minor difference between the just-mentioned periods. The authors think that these results indicate the prevalence of TB in ancient Egypt. They presume that the presence of mTB in normal bones was due to antemortem, systemic spread, and generalized infections. The authors state: "the extent of mycobacterial infection and possible differences in its spread during the three-thousand-year history of ancient Egyptian history remains unclear at present".

Zink et al. in a previous paper found comparable results. They state that no data are available yet on the frequency of infectious diseases in previous times.

As mentioned previously, the living conditions and the weather in ancient Egypt did not promote endemic TB.

Ziskind and Halioua in a recent publication concluded that TB certainly plagued the Nile Valley and appears to have been an important cause of mortality in ancient Egypt. They base their conclusions on molecular biology and identifying TB by use of polymerase chain reaction (PCR), which was positive in a third of Egyptian mummies. Spoligotyping made it possible to re-evaluate the phylogenic type of the mycobacterium TB complex in the ancient Egyptians.

4. Orthopedic pathology:

   a. In a recent study by Nerlich et al., they found:

      1. Bone TB.

      2. Chronic anemia (porotic hyperostosis, cribra orbitalia).

      3. Degenerative disease of the bone (possibly Paget's disease).

      4. Degenerative diseases of the joints and the spine.

      5. Metabolic disease of the bone.

      6. Metastatic bone disease.

      7. Osteomalacia.

      8. Osteomyelitis.

      9. Scurvy.

     10. Trauma in the bones.

b. From the mummies in the London Museum, the following bone pathologies were detailed:

1. Dental caries.

2. Dislocation of the jaw.

3. Dwarfism (see Nunn, p. 78).

4. Hydrocephalus (Derry, 1942).

5. Mastoiditis.

6. Osteogenesis imperfecta.

7. Osteoporosis.

8. Scoliosis.

9. Skull fractures.

10. Talipis equinovarus (possible postpoliomyelitis).

11. Paleo-oncology.

Cancer in ancient Egypt was probably a rare disease due to the fact that life expectancy was not very long. Environmental pollution with toxic gases and other materials did not exist and of course tobacco smoking was unknown to the ancient Egyptians. Elliot Smith and Wood Jones with their extensive work at Aswan before building the first dam found no bony evidence of metastatic cancer. Elliot Smith and Dawson were unable to report any cases of cancer from the Pharaonic period. In 1986, Ghalioungui reviewed scanty cases. Strauhal et al. reported a case of carcinoma of the nasopharynx from Egyptian skeletal remains from a cemetery at Naga-ed-Dêr. They quoted three other cases reported by Derry in a skull found in Nubia, another case was reported by Wells, and more recently a case was reported by El-Rakhawy et al. A case of metastasis reported by Nerlich evidence of cancer in the medical papyri is very uncertain. Khonsu's tumors were related to leprosy (see Eber's #813). Cancer of internal organs has not been reported. Cancer was, therefore, rare in ancient Egypt. Perhaps cancer of the bladder due to bilharziasis might have existed, but not documented. Cancer of the breast was not reported.

CHAPTER 11

Epidemiology of Cardiovascular Diseases (in the Modern Era)

ACCORDING TO THE WORLD Health Organization estimates, 16.6 million people around the globe die of cardiovascular diseases (CVDs) each year. (See statistics from more recent data by the American Heart Association, 2009.)

In 2006, CVDs accounted for 198,000 deaths in the United Kingdom (UK). CVD causes over 4.3 million deaths in Europe and over 2 million deaths in the European Union (EU). It accounts for 48% of all deaths in Europe and the EU.

Coronary heart disease is the single-most common cause of death in Europe, accounting for 1.92 million deaths each year

In 2005, stroke caused 5.7 million deaths worldwide. Stroke is the second-most common cause of death in the EU and the third leading cause of death in Canada. In the UK, about 68,000 new cases of heart failure occur each year. In the UK, 31% of men and 28% of women have hypertension. Congenital heart disease is associated with about 4,600 babies born each year in the UK. An estimated 12 million have rheumatic fever and rheumatic heart disease worldwide. An additional 27 million people are estimated to have peripheral vascular disease in Europe and North America.

Tobacco use studies in 2006 found that 23% of men and 21% of women of all ages smoked cigarettes. The mortality from coronary heart disease was 60% higher in smokers and 80% percent higher in heavy smokers. According to the CDC and the WHO, smoking results in a 100% increased risk of stroke and coronary heart disease, more than a 300% increased risk of peripheral artery disease, and a 400% increased risk of aortic aneurysm. Smoking kills over 1 million people in Europe each year.

Overweight and obesity have tripled since the mid-1980s. The WHO estimates that more than 1.9 billion adults were overweight and more than 650 million adults were obese by 2016 worldwide. In Canada, almost 60% of people are overweight or obese.

In the UK, 5% of men and 4% of women are diabetic. This trend is rising in children and adolescents.

For detailed statistics, please see the American Heart Association's *International Cardiovascular Disease Statistics*.

For further detailed information on the United States statistics: "Heart Disease and Stroke Statistics 2009 Update: A report from the American Heart Association Statistics Committee and Stroke Subcommittee". Circulation 2009; 119; e21–e181; originally published online December 15, 2008; DOI: 10.1161/CIRCULATIONAHA.108.191261.

Websites for further statistical analysis and information on CVD:

1. United Nations Population Fund: www.unfpa.org

2. World Health Organization Statistical Information System (WHOSIS): www.who.int/whosis/en

3. British Heart Foundation—Coronary Heart Disease Statistics: www.heartstats.org/homepage.asp

4. European Society of Cardiology: www.escardio.org

5. G8 Promoting Heart Health: www.med.mun.ca/g8heart health/pages/enter.htm

Clinical Aspects of Cardiovascular Diseases in Ancient Egypt (From the Ebers Papyrus: The Vascular Text)

I N THIS CONCISE PRESENTATION, cardiovascular medicine in ancient Egypt will be summarized, including aspects of anatomy, physiology, pathophysiology, clinical issues, and paleopathology. This information is based on the vascular book of the Ebers Papyrus and, to a lesser extent, on the Berlin Papyrus and Edwin Smith Papyrus. In ancient Egyptian medicine, the heart is known by two names, *ib* and *hatty*, which are used interchangeably. The word *ib*, but not *hatty*, is used for the heart as the center of emotions as in other languages. The stomach is *r ib*, which means the mouth of the heart, which is really the mouth of stomach, i.e., the cardia of the stomach. (We still use "heartburn" for acid stomach.) Another clarification is about the vessels (the *metu*, which is the plural of *met*): this term is used for cordlike structures whether they are rather solid tendons or a hollow tube containing blood (or air), namely, the blood vessels or those containing urine, such as ureters.

The ancient Egyptians knew the heart and the vessels. The physicians and the embalmers knew the heart's position. The physicians knew that it is a central pump, pushing blood through the *metu* and felt as the pulse (the heart speaking).

They had no watches to count the pulse, but they compared it with the pulse of the examiner. They assayed its strength, weakness, and irregularity, and could use it to predict cardiac disease. Halioua and Ziskind have analyzed the book of the heart, from where it is stated: "In what is called the dance of the heart 'hatty' the patient's heart is observed to move away from the left breast" – a displacement that is today diagnosed as tachycardia or dysrhythmia due to increased cardiac volume (Ebers 855n).

Weakness of the heart was observed by the Egyptian physician. It was noted that weakness of the cardiac pulsation was followed by bodily weakness (Ebers 855q).

Angina pectoris was classically described as chest pain and pain in the left shoulder, left arm, and left side of the stomach; it has little chance of recovery (myocardial infarction). Myocardial infarction was clearly described as weakness of the cardiac pulsations, displacement, and generalized changes (cardiogenic shock). The Egyptian physician recognized the seriousness of these signs in their patients (Ebers 855q).

Heart failure was described in Ebers 855d, in which case the pulse is weak, the liver is enlarged, the lungs are congested, and the extremities are cold.

With the weakness of the heart, there is reduction of the blood supply to the brain leading to lethargy. The heart no longer speaks, and the vessels become "dumb" under the fingers of the doctor (Ebers 855e).

According to Ebers 855p, when the heart is displaced, it is a sign of disease.

From the heart, 22 pairs of vessels go peripherally, classified by Nunn. Those going to the limbs match our modern knowledge.

The earliest description of cyanosis, a cardinal clinical sign of congenital heart disease, was given by the ancient Egyptians. The earliest illustration of a blue-skinned person was taken from the tomb of Osiris, dating back to 2000 B.C. The Edwin Smith Papyrus, according to Willerson and Teaff, presented convincing evidence that cyanosis was diagnosed even earlier than 2000 B.C. The greater Berlin Papyrus or the Brugsch contained a description of the heart similar to that in the Ebers Papyrus. It also presented a superficial description of the anatomy of the veins. Some historians believe that this papyrus is referenced in Galen's writings.

VASCULAR TUMORS

There is a description of a blood vessel tumor. The tumor was described in the Ebers Papyrus as coming from the vessel after being injured. The treatment was removal of the tumor by a knife that had been heated in fire to reduce bleeding (compare with the Bovie method of the present day).

VENTRICULAR FIBRILLATION

In the Ebers Papyrus, as long as the heart is strong, the pulse is readily palpated; when the heart is weak, the pulse may not be felt. When the heart trembles, it has little power and sinks. The disease is advancing: the trembling of the heart most probably is due to ventricular fibrillation. A dropped heart beat was described as the forgotten heart.

Peripheral vascular diseases were described as numb vessels or the pump being weak or not felt. Varicose veins, arteriovenous malformation, and traumatic arterial aneurysms were also described.

Enlarged liver with marked congestion was described, probably Egyptian hepatosplenomegaly due to schistosomiasis.

KNOWLEDGE OF THE CIRCULATION

The scientific pattern of the circulation was not known to the ancient Egyptians, but they had practical observations about it.

VENOUS RETURN

As regards the venous return, we have an example from the Carlsberg Papyrus: if you put a clove of garlic in a woman's womb all night, and if she smells garlic in her breath in the morning she can conceive. Thus the course of the venous circulation to the right ventricle, the pulmonary artery and then the lungs was recognized.

Pathological conditions of the heart summarized in the Ebers Papyrus 855b include:

1. Description of heart failure: weakness of the heart and it is flooded (filled mouth with fluid due to pulmonary edema).

2. *Angina pectoris* was typically discovered and in several cases of myocardial infarction, it was stated that the patient is approaching death.

3. Cardiac enlargement due to failure or other causes.

4. Cyanosis due to congenital heart disease.

In peripheral *metu*:

1. Dumb vessels due to disease of the vessels or weakness of the heart (normally carry air to the tissue).

2. Traumatic arterial aneurysm.

3. Varicose veins, AV malformation.

4. Possible portal hypertension (hepatomegaly).

The cardiovascular aspects are extracted from the Ebers Papyrus #854 and #856 and the Berlin Papyrus #163. They have been classified in detail by Nunn, as well as tabulated.

Out of the 829 prescriptions in the Ebers Papyrus, the following are applicable to cardiovascular diseases. They will be numbered in a series, rather than having their corresponding numbers in the papyrus. Comments have been added to clarify the statements:

The beginning of the physician's secret: knowledge of the heart and its movements.

1. There are vessels from it to every limb. As to this, when any physician, any surgeon (Lit. *Sakhment*-priest), or any exorcist applies his hands or his fingers to the head, to the back of the head, to the hands, to the place of the stomach (epigastrium) to the arms or to the feet, then he examines the heart, because all his limbs possess its vessels, that is: it (the heart) speaks out to the vessels of every limb.

   Comment: This is a very clear description of the pulse and that its origin is the heart beat transmitted along the arteries, i.e. an idea about the arterial system (cf. Doby, 1963; Kirchner, 1878; Piankoff, 1930).

2. There are four vessels in his nostrils: two give mucus and two give blood.

   Comment: This referred to the veins whose congestion produces more mucus and the arteries whose injury give rise to bleeding (epistaxis).

3. There are four vessels in the interior of his temples, which then give blood (to) the eyes; all diseases of the eyes arise through them, because there is an opening to the eyes. As for the water (tears) that comes down from them, it is the pupils of the eyes that produce it; another reading: it is the sleep in the eyes, which makes it.

Comment: When the pupils produce the tears, it means a reflex phenomenon, which occurs in iritis or photophobia in eye inflammations. Eye diseases may be associated with temporal arteritis (Rowland, 2000).

4. There are four vessels dispersing to the head, which effuse in the back of the head and which then produce a bald spot? And loss of hair (?); this is their production upwards.

Comment: The writer describes occipital vessels. Prolonged recumbency in the supine position, especially due to paralysis produces alopecia areata due to pressure ischemia, which may even produce *occiput decubitus* ulcers.

5. As to the breath, which enters into the nose, it enters into the heart and the lung; this gives air to the whole belly.

Comment: This suggests that the ancient Egyptians hypothesized about the lesser and greater circulation, where air (O_2) passes to the lungs and to the heart and from there goes peripherally, especially to the abdomen. They probably knew it passed into the aorta, which terminates in the abdomen, and they detected in the embalmment.

6. As to that through which the ears become deaf, there are two vessels that affect it (namely) – the ones leading to the root of the eye; another reading: to the whole eye. When he is deaf, his mouth cannot be opened (i.e. he cannot speak).

Comment: The vessels mentioned here may be the eustachian tubes. They probably meant trismus, which occurs with otitis media with some hearing loss.

7. Another reading: as to that through which the ears become deaf; it is these "vessels" are on the temples of a man *h r nssw*; it is these (vessels), which give a cutter in a man, so that he (the cutter) takes for him his hair? As to the inundation of the stomach, it is (due to) fluid of the mouth, all his limbs become faint.

Comment: The writer means fainting when sialorrhea occurs and the heart becomes weak; i.e. syncope.

8. As to "SS of the stomach" it is a vessel whose name is receiving (SSPW), which makes it, it gives humor to the heart, and all his limbs become weary after yonder receiving by the heart thereof.

Comment: This refers to the aorta from the heart supplying all the limbs and giving humor to the heart; probably referring to the coronary arteries. It was not clear how the heart, which was considered to be the "source of blood", supplied itself with blood.

9. As to "debility, which arises in the heart", it is *h 3 s f* as far as the lung and liver, deafness (insensibility?) comes forth for it (i.e. the heart) and its vessels fall down after their beat. *h 3 s f s f h – b r.*

Comment: This was probably a picture of congestive heart failure with congestion of the lungs and liver.

10. There are four vessels to his two ears together with the (ear) canal, (namely), two on his right side and two on his left side. The breath of life enters with the left ear; another reading: it (i.e. the breath of life) enters into the right side and the breath of death enters into the left side.

Comment: It is difficult to explain this because it may be a superstition. It is still believed that items on the left imply pessimism and vice-versa. In Egyptian villages, until now it is believed that one's soul after death remains in the left ear until the body is buried. Even final prayers are whispered to the dead in the left ear.

11. There are six vessels that lead to the arms: three to the right, and three to the left, they lead to his fingers. There are six vessels that lead to the feet, three to the right foot and three to the left foot until they reach the sole of the foot.

Comment: This is an anatomical description of the arteries in the arm, where the brachial artery divides into the radial and ulnar arteries and in the lower limbs where the popliteal artery branches into two tibials.

12. There are two vessels to his testicles; it is they that give semen.

Comment: The writer meant the testicular arteries, and he observed the physiological function of the testicle producing semen (sperm).

13. There are two vessels to the buttocks, one to the right buttock and the other to the left buttock.

Comment: The writer meant the internal iliac vessels with their gluteal branches.

14. There are four vessels to the lung and to the spleen; it is they that give humor and air to it likewise.

Comment: The writer meant the pulmonary veins, and it seems that all the vessels he mentions are those related to the arterial blood. "Giving humor and air" meant oxygenated blood coming by the pulmonary veins. Mentioning the lungs together with the spleen may be based on the similarity of appearance of both organs to the naked eye.

15. There are four vessels to the liver, it is they which give to it humor and air, which afterwards cause all diseases to a rise in it by overfilling with blood.

Comment: This description entails both portal branches bringing humor (nourishment) to the liver and the hepatic artery branches giving air or oxygenated blood. Metu is the bile ducts forming the common bile duct. What is probably meant by "overfilling with blood" in disease condition is portal hypertension. This could be the first description in the world of portal hypertension, commonly due to schistosomiasis in Egypt, and leading to cirrhosis of the liver.

16. There are two vessels to the bladder; it is they, which give urine.

    Comment: Here, the ureters are meant as the vessels to the bladder bringing urine down from the kidneys (the *metu*).

17. There are four vessels that open to the anus (rectum?); it is they, which cause humor and air to be produced for it. Now the anus opens to every vessel to the right side and the left side in arms and legs when (it) is overfilled with excrements.

    Comment: This describes the rectal blood supply. Arms and legs meant ascending and descending branches. This is probably a description of congestion of the rectum in constipation or impaction with stools (dyschesia). Studies were done by Chapelain and Jaures (1920).

18. As to "faintness" it is (due to the fact) that the heart does not speak or that vessels of the heart are dumb, there being no perception of them under the fingers (i.e. thou doest not feel them); it arises through the "air" which fills them.

    Comment: This is a description of impalpable or very weak pulse in the states of collapse or shock and with probable hypoxia. It most likely meant the lack of the "air" (oxygenated blood) due to hypotension (possibly describing cardiogenic shock?).

19. As to the "feeling of sickness": it is (due to) debility of the heart through heat from the anus if thou findest it (the sickness?) great, something ꜥ h p (rotates?) in his cardia, likewise in the eye.

    Comment: "Feeling of sickness" may refer to fainting spells, which may occur in habitually constipated individuals with intestinal intoxication. The reference to cardia may mean nausea. Vision impairment and eye rotation may occur. This is a common syndrome seen in Egyptian peasants.

20. As to "his mind (consciousness?) passes away": it is due to the fact that the vessels of the heart are carrying "feces".

    Comment: The same condition as above with skatole and indole odor carried to his breath or cases of intestinal intoxication with fetor oris.

21. As to "all dropsical diseases that enter into the left eye and go forth from pudenda": it is (due to) the breath of the activity of the priest. It is the heart which causes them to enter into his vessels, and it boils and boils in all his flesh, the heart "n d h d h" to him through them, because it comes boiling and the vessels of his heart become faint to him thereby.

    Comment: This may refer to an infectious fever of unknown nature, while "boiling" may mean rigors accompanying it (possibly septicemia).

22. As regards the phrase "they displaced (?) their clothes": it means dropsical diseases, increasing body size. As to "his dropsical diseases are high": this means that they overflow.

    Comment: This may be a description of ascites or general anasarca, where the body is edematous, displacing the patient's clothes.

23. As to the "mind's kneeling (breakdown)": this means that his mind is constricted and his heart in its place and the blood of the lung becomes small through it. It is due to the fact that the heart is hot, and then his mind becomes weary through it; he eats little and is fastidious.

    Comment: The writer may mean a heart attack, where the patient may be obtunded and/or incoherent.

24. As to the "drying up of the mind": it is (due to the fact) that the blood *d m 3* (coagulates?) in the heart.

    Comment: This may mean heart attack with mental confusion.

25. As to the "mind kneels through purulency": this means that his mind is small in the interior of his belly, the purulency falling on his heart and so he becomes *I 3 r* and kneels.

    Comment: Purulency means here decay or senility or dementia possibly due to arteriosclerosis.

26. As to the "debility through senile decay": it is (due to the fact) that purulency is on his heart.

    Comment: By "purulency on the heart" the writer meant senile debility, which affects many organs, including the heart.

27. As to the "heart dancing": this means that it moves itself to his left breast (mamma) and so it pushes on its seat and moves from its place; this (i.e. phrase "its place") means that its adipose sac (?) is in his left side towards the joining with his shoulder.

    Comment: This most probably meant enlarged hearts complicating rheumatic valvular diseases with precordial pulsations. By "dancing" the scribe probably meant atrial fibrillation, which may accompany this cardiac condition.

28. As to "his stomach is very (too?) low": this means that his stomach has sunk, it having proceeded downwards, and it is not in its right place.

    As to "his heart in its (right) place": this means that the adipose sac (?) of the heart is in his left side; it does not go upwards and does not fall downward through anything, remaining in its place.

Comment: The low stomach probably was due to gastroptosis. By "the adipose sac of the heart", the writer meant the pericardium with the overlying fat. These are two unrelated conditions.

29. As to his heart *n b 3* (?) – *f c 3*, the adipose sac (?) under his left breast (*mamma*): this means (or it is due to the fact) that his heart has made a little going downwards, and so his disease passes away (?).

    Comment: This could mean a pericardial effusion, but it is unclear.

30. As to "his cardia" *h s f*, this means that his cardia is big. As to the mouth is hot and *h n w s* (stings?) and as to the "stomach *h n w s*": it is due to the fact that heat has passed over his heart, and that his stomach is hot on account of burning, as (in) a man whom *h n w s* has heard (?).

31. As to the "rotation falls on his heart": this means that heart rotation falls on his heart, that may become faint, and that his mind is consumed through *d n d*. It is his heart's overfilling with blood, which it does it, (an overfilling) that arises through drinking of water and eating of hot *s h j t* – fishes, it causes (it) to arise.

    Comment: These may be cases of congestive heart failure, which is aggravated by drinking much water and salted fish.

32. The beginning of the book on the traversing of purulency in all limbs of a man according to what was found in writing under Anubis' feet in Letopolis; it was brought to his Majesty of upper and lower Egypt Usaphia, the justified.

    As to man, there are 22 vessels in him to his heart; they give to all his limbs.

    There are two vessels in him in *s r t j w* (the superficial venous plexus?) to his breast (*mamma*), they make burning in the anus. What is done against it: fresh dates, *h m w* of *k 3 t p 3 w t* of sycamore, are pounded together with water strained and let be taken four days.

Comment: This could be a description of veins. However, this article is hard to comprehend.

33. There are two vessels in him to his thigh, if he is ill in his thigh or his feet, ache (?) then thou shalt say concerning it; it is (due to the fact) that the vessel *s r t j w* (the superficial venous plexus?) of his thigh, has received the illness. What is done against it: viscous fluid, *S 3 M*, natron, are boiled together and drunk by the man for four days.

Comment: This may be a description of varicose veins. The boiled natron may be acting as a diuretic or laxative to relieve the congestion.

34. There are two vessels in him to his nape, if he is ill in his nape and his eyes are dim-sighted, then thou shalt say concerning it: it is (due to the fact) that the vessels of his nape have received the disease. What is done against it: myrtle (?) Washer-man's slops, pignon, fruit of a *3 m s*, are mixed with honey, applied to his nape and (it) is bandaged there it for four days.

Comment: Probably refers to occipital neuralgia or occipital headache.

35. There are two vessels in him to his arm if he is ill in his shoulder or his fingers ache (?) then thou shalt say concerning it: it is a "rheumatic" pain. What is done against it: let him vomit by means of fish with beer and *d 3 j s* or meat and his fingers are bandaged with watermelon, until he is healed.

There are two vessels in him to the back of his head.

There are two vessels in him to his forehead.

There are two vessels in him to his eyebrow.

There are two vessels in him to his nose.

There are two vessels in him to his right ear; the breath of life enters into them.

There are two vessels in him to his left ear; the breath of death enters into them.

Comment: This is rather an inaccurate anatomical description. The cause of death through the left ear is unknown. It may relate to vagal inhibition reflex? It could be a belief since they knew the heart is on the left side? It is hard to interpret.

36. All together (they) go to his heart, divide to his nose, all together (they) unite to his hinder parts, and illnesses of the hinder parts arise through them; it is excrements that are carried, it is the vessels of the feet that begin to die.

    Comment: Here the venous circulation may refer to the vena cava caudalis, but this is only a presumption.

    In the surgical section among vascular conditions, it is stated:

37. If thou examinest a swelling of the covering on his belly's horns above his pudenda, then thou shalt place the finger on it and examine his belly and knock on the fingers; if thou examinest this? that has come out and has arisen by his cough, then thou shalt say concerning it: it is a swelling of the covering of his belly, it is a disease that I treat. It is heat on the bladder in front in his belly, which causes it (i.e. the covering) to fall downwards, and the return is likewise. Thou shalt heat it (i.e. the swelling) in order to shut (it) up in his belly; thou shalt treat it, as *s 3 h m m* treats.

    Comment: This is a case of hernia, probably associated with micturation disturbances, or possibly due to an enlarged prostate.

    Instructions concerning a swelling of the lower part of his belly:

38. If thou examinest this on the lower part of his belly, water in his belly moving up and down, then thou shalt say concerning it: (It is) an affection h r w – t 3 w in the lower part of his belly; it is a disease which I treat. It is heat on the bladder that causes it. Thou shalt hit it (i.e. the swelling) into him with a h m m instrument not descending into his navel. Thou shalt treat him as s 3 h m m treats.

Comment: The papyrus describes ascites (?) and a full bladder with secondary inguinal hernias. By "heat on the bladder" it meant oliguria (hot or dry bladder), which can occur with ascites. He advised only reduction of the hernia (thou shalt heat it (i.e. the swelling) in order to shut (it) up in his belly). The reduction of the hernia was probably performed after seating the patient in a warm sitz bath.

Instruction concerning a swelling of vessels:

39. If thou examinest a swelling of a vessel in any limb of a man, and thou findest that it is hemispherical (?) and grows under thy fingers on every going (i.e. pulsation of the heart), (but) if it is separated from his body, it cannot on account of that become big (i.e. grow) and not give out (i.e. diminish) then thou shalt say concerning it: it is a swelling of a vessel; it is a disease which I will treat: it is vessels that cause it, and it arises through injury to a vessel. Thou shalt perform an operation for it, heat with fire, it shall not bleed much. Thou shalt treat it as *s 3 h m m* treats.

Comment: This is descriptive of an arterial aneurysm, probably post-traumatic. He described its pulsations and that pressure on the vessel above the swelling would cause it to become smaller, as if it were separated from the body. The treatment by red-hot fire would cause thrombosis.

40. If thou examinest a swelling of vessels on the leather layers (i.e. of the cutis) of any limbs, and its appearance is growing

on account of serpentining of the serpentry and they (i.e. the vessels) have formed many knots, that which is like something inflated with air, then thou shalt say concerning it: it is swelling of vessels. Thou shalt not put thy hand to such a thing; it is that hurts the limb in his arm. Thou shalt prepare (the healing of vessels in all limbs of a man). Which *s r t j w* (the superficial venous plexus?) Which *s r t j w* me, and which jumps (i.e. pulsates) in the midst of these limbs because thou unitest with Chons' unions. If thou examinest Chons' swelling *m 3 ᶜ n d m n b k w j.* Let me bring artificial gifts to Re, namely faience, in the mornings. Is recited four times very early in the morning.

Comment: This is a description of an arterio-venous or cirsoid aneurysm or fistula with its serpentine secondary varicose veins and its transmitted arterial pulsations to the venous side. He does not advise operation on it probably because of danger of bleeding and recurrences, which do occur, especially in the congenital form. Therefore it was left for Re and recitals for treatment by magic.*

Instruction concerning oozing in any limb:

41. If thou examinest an oozing vessel in any limb and thou findest it blush red and hemispherical (?) from the blow of a stick or from the stroke of any thing to any limbs, after he has made seven knots, then thou shalt say: it is an oozing from a vessel; it is an injury to a vessel which causes it. Thou shalt perform an operation for it with *s w t* – reed, (such as is used) for making operations. If it bleeds much, thou shalt burn it with fire, thou shalt treat it with *s 3 h m m*'s treatment.

---

* *Khonsu's Tumor is a swelling of limbs; may be lepromatous or gangrene (see Ghalioungui, 1936: p. 87).*

Comment: This is a description of blunt vascular injury, particularly to a vein, which leads to a hematoma. If it was large, he advised evacuation, and in case bleeding persisted, he advised fire coagulation.

42. If thou findest on the leather layers (i.e. the skin) of any limb as many serpentine windings, which are inflated with its air, then thou shalt say it is an enemy of the vessel. Thou shalt not put thy hand to any likeness of this that should be upside down.

Comment: This referred probably to varicose veins distended with blood (air). Operation was not advised, but elevation of the limb was advised.

43. The beginning of remedies to treat the right side in hemiplegia(?): fresh porridge *1 r c*, mustard (oil) *2 ro, d s r t* – beer, (It) is bandaged with.

Another: Frankincense ½ *ro*, fruit of juniperus 2 *ro*, 1 *b w* from Lower Egypt 2 *ro, i b b j* 2 *ro*, celery from the hill country 2 *ro*, northern celery 2 *ro, p s n t* 2 *ro, t jᶜ 3 m* 2 *ro, s w t* 2 *ro, b b w* 2 *ro, s w t d h w t j* 2 *ro*, white mustard (oil) 2 *ro*, green mustard (oil?) 2 *ro*, pinetar 5 *ro, 9 3 j t* 2 *ro*, hyoscyamus 5 *ro, r w d* 2 *ro*, myrtle (?) 4 *ro*, honey 1 *ro*, (it is bandaged with).

Another: Saffron 1, raisin 1, ammi 1, porridge 1, *h m 3 w* l, Costus (?) 1, bran (?) of barley, are mixed together and the side is bandaged therewith.

Comment: This is an example of phytotherapy to induce hyperaemia in the affected limb.

44. The beginning of remedies against fetid nose (ozena): date wine, its opening is filled therewith.

Comment: This is an example of using wine to excite local hyperemia and disinfect the unhealthy mucosa, which is atrophic.

45. If thou examinest a man with a resistance in his left side and it is under his flank and does not cross the earth (stretch across the abdomen?), then thou shalt say of him: it has produced like a shore and formed a *s ͨ j t* – cake. Thou shalt prepare for his remedies (for) within him in front: ammi 8 *ro*, ground *t j m* 4 *ro*, pignon 2 *ro*, *s 3 s 3* 4 ro, are boiled together with oil 2/3 and honey 1/3, are eaten by the man for four days.

    If thou examinest the man after this has been done, and thou findest that it spreads and goes downwards, then thou shalt prepare for him: powder of hyoscyanus, is boiled all through and eaten by the man to fill his belly and *s p 3* his intestines for four days. Thou shalt lay thy hand on him; if thou findest it cut and destroyed, as if it was grain, then thou shalt prepare for him berry-juice to cool: ammi 1, *i w b 1*, water, are strained and taken for 4 days.

    Comment: Here the resistance on the left side of the belly, its edge shore like (the edge with notches) may be due to malarial or bilharzial spleen (Egyptian splenomegaly).

46. If thou examinest a man for illness in his cardia, and he has pain in his arm, in his breast (mamma), and in one side of his cardia, and it is said of him: it is *w 3 d* – illness, then thou shalt say thereof: it is (due to) something entering into the mouth, it is death that threatens him. Thou shalt prepare for him stimulating herbal remedies: fruit of *t h w 3 1*, *h 3 s j t* 1, *n j w j* 1, thyme (?), red grains of mustard seed 1, are boiled with oil and drunk by the man. Thou shalt apply the extended (i.e. flat) hand on him until the arm gets well and rid of pains. Then thou shalt say: this suffering has been descended to the real intestine (rectum?) and to the anus; I do not at all repeat the remedy.

    Comment: This was a description of angina pectoris with its radiating pain. The remedies contained vasodilators, which may improve the condition. If the condition worsened, the patient may develop gastrointestinal dilatation (ileus), which was terminal and could not be treated by the physician.

Section 99–103, 18, which contains information of the vessels of the body as well as the so-called "glosses", deserves special comment.

In the first passage, it is said that the pulsation or the "speaking" of the heart may be felt through these vessels. It is further stated that they provide the substance that is secreted from the nose, the testicles, and the urinary bladder, and also bring humor and air to some of the internal viscera (the liver, lung, spleen, and gut). In the second passage, the vessels are described as the canals through which the disease spreads in the body, the vessels are said to have received (and transmitted) the illness in question. Evidently there are two different things referred to, viz.: The arteries that transmit the pulsation or the "speaking" of the heart and the air to the viscera (the oxygen) and the veins that transmit the disease and spread it into the body by going through the heart.

In the first "gloss" (99, 12–14): "As to the breath which enters into the nose", it enters into the heart and the lung. This indeed is a statement about the further passage of the breath, after it has passed through the nose.

This suggests an understanding of the lesser circulation as imagined by the ancient Egyptian physicians.

The Egyptians recognized the heart as the center of the vascular system and believed it to be connected with other organs of the body by a system of 24–36 vessels extending down to the little fingers – hence the custom of dipping the little fingers during the ceremony of initiation.

According to a Memphite theological treatise, "the action of the arms, the movement of the legs and of every part of the body is dictated by order of the heart". All the senses function from it: "the sight of the eyes, the hearing of the ears, the breathing of air through the nose are all controlled by the heart. It is the heart, which decides and the tongue proclaims what the heart has conceived". Wear and tear on this essential organ brought about senility. Describing old age Sinuhe writes: "My eyes are heavy, my arms without strength, my legs refuse to work because my heart is tired".

THE ROLE OF THE HEART IN EGYPTIAN THOUGHT (FROM *DAS HERTZ IM UMKREIS DES GLAUBENS* BY BRUNNER, 1965)

The Egyptians understood the physical functions of the heart; and attributed nonphysical functions to it as well.

In everyday life, the heart was considered as the seat of:

1. Wisdom and knowledge. Thus the early Egyptian thought with his heart. His limbs merely carried out its commands.

2. Memory, as expressed in the English phrase "to learn by heart".

3. Sentiment and feelings, which is similar to our own ideas.

4. Worry, fear, and happiness.

5. Love, which is also in line with our own beliefs.

6. Compassion.

7. Cunning and malevolence.

8. Desire: the man did not wish for anything—his heart wished for him. cf. English: to gain one's heart's desire.

The word *abi* signifies longing, desire, will, wisdom, courage, etc. The heart was regarded as the seat of life feelings and of all its activities both moral and physical. There was also a soul (*hati*) connected with the heart; it is the source of its activities and mental condition i.e. the soul of the heart. The *abi* and the *hati* could be stolen, which would result in death. The Egyptians recognized the influence of emotions such as grief, joy, desire, stress, on the heart "center". It was considered to be the site for such feelings and from it a sort of "spirit" or humor affected the general condition of the individual. In many inscriptions, brave people were described as having hearts of bronze. The heart expressed honesty and dishonesty, good character, love, which is similar to contemporary expressions. The ancient Egyptians, as well as the Hebrews and Arabs, considered the heart the center of memory, thinking, wisdom, intelligence, and spiritual affections.

The heart was thus believed to be the center of physical and emotional life, of the will and the intellect. All the feelings, conditions of the soul, traits of character, and temperament were expressed in Egyptian by various idioms referring to the heart. "Happy" was described as "long of heart"; "depressed" was "short of heart"; "attentive" was "counted of heart". The confident was called "he who fills the heart". "To drown the heart" meant "to hide one's thoughts"; "to wash the heart" was "to satisfy a desire". A scholar has collected 350 expressions of this kind without completing the possible list. The hieroglyph heart appears continually in the texts.

In the *Book of the Dead*, as well as other references, the relation of the heart to the spirit after death is very close; one can find a very rich literature, especially the *Book of the Dead*. A translation of the heart in the belief is summarized in the following in H. Brunner's *Das Hertz im Umkreis des Glaubens* (1965).

The heart was of great importance in religion. According to the Memphite cosmogony, the god Ptah conceived of the world in his heart before bringing it into existence through his creative utterance.

The heart was the conscience; it dictated a person's actions and rebuked him; it was an independent being of a superior essence dwelling in the body. Written on a coffin in the Vienna Museum are the words: "A man's heart is his own god". To the Egyptians the heart was the center of all activities, not the brain, which was a servant to the heart. ·

The religious implications of the concept of the heart as the seat of understanding and wisdom are to be found in the *Ma'at*. The *Ma'at* was the world as God made it, and its development under the influence of man. The aim of the *Ma'at* was to outline the correct relationship of man to the situations with which he was faced. The wise men who understood the *Ma'at* taught their findings to their pupils. These wise men had been given their hearts so that they could transmit the *Ma'at* to others.

The heart was regarded as a gift from "God". Thus if the Egyptians followed the commands of their heart, they believed that they would be successful. As the organ that recognized the will of "God", the heart had to be "firm", i.e. it could not waver from carrying out "God's" will. This led the Egyptians to believe that the heart had a will of its own. It could command man to do this for pure enjoyment. This they justified on the basis of the belief that you cannot use it when you die. Egyptian religion was thus not entirely dominated by impending death.

The heart played a central role in the Last Judgment. The Egyptians believed in life after death. Before this second eternal life, they were judged on what they had done on Earth. The deceased had to confess all his sins. He could not lie, as his heart could speak against him, if it chose to. This proved a drawback in the Egyptians' religion, and they took great pains to ensure that their hearts would not speak against them.

The judges in the other world caused the dead person's heart to be weighed (v. weighing the heart) to decide whether by his conduct on Earth, he deserved the immortality of the blessed. For this test, the heart was dissociated from its owner, it became a stern witness, and the owner begged it to give a favorable report. In the mummy, the heart was accompanied by a funerary amulet (the heart scarab) on its flat surface, which carried a message from the Chapter XXX of the *Book of the Dead*, which states,

> O my heart don't witness against me, don't accuse me in front of the tribunal, don't be against me in the presence of the balance (scale) carrier Anubis, say nothing about my lies against me in the presence of the Great God of the West (Osiris) (see Figure 12.1).

The embalmers, who removed most of the viscera from the body, left the heart in position in the mummy believing that it was required to undergo the last judgment in the afterlife. The *Book of the Dead* contained spells for restoring the heart to the dead person in the afterlife.

It was of prime importance to the Egyptians that the heart should not leave the body. From a technical point of view, it was,

FIGURE 12.1   Funerary Amulets.

however, very difficult in the early days to keep the heart when embalming the body. They had to remove the heart with the other viscera from the body, but they kept it separate from them. Although the heart was laid to rest with the body, they greatly feared that the two would be parted. The records say that they put imitation stone hearts into the body so that the deceased would always have a heart, but no examples have been found. From c. 1500 B.C., however, they could embalm the heart within the body, but there was still no assurance that the heart would corroborate the deceased's confession. So a stone heart with an inscription on it was strapped to the left breast of the deceased. The inscription asked the heart not to denounce any claims made at the Last Judgment.

Finally, the Egyptians believed that each individual had his own god, whose place was in the heart. Thus the heart was regarded as man's god or master. The heart was an inner partner with its independent opinion. This was their own personal god, but at one and the same time, they had one great God for the entire world.

It is interesting to compare the ancient Egyptian views on the heart with the views of other civilizations.

The Babylonians believed that the liver was the seat of the soul or life, and it was only later that they recognized the importance of the heart. Many traces of the earlier idea survived, "the liver" being used instead of, or along with the heart, to signify the seat of life, and hepatoscopy was constantly practiced. A frequent refrain in the hymns is: "May the heart be at rest, the liver be appeased".

Among the ancient Hebrews, traces of the same idea of the liver being the seat of life and emotion are to be found. However, later in the Old Testament, the heart is considered the source and symbol of life. This is mentioned in Proverbs 4:23: "Keep thy heart with all diligence, for out of it are the sources of life". The heart is the seat of intellectual (to the exclusion of the head) and emotional life, including thought, memory, perception, will, imagination, joy, sorrow, anger, etc. (Psalms 33:11, Psalms 104:15, Genesis 8:21, Deuteronomy 29:4, Deuteronomy 4:9, Deuteronomy 19:16,

Exodus 14:5, 1 Kings 8:38, Isaiah 30:29, and Isaiah 66:14). Hence, such phrases as men of heart (= men of understanding) Job 34:10, out of heart (forgotten) Psalms 31:12. The heart is also the seat of religious feeling, hence, "change of heart", a "clean" heart; and a "heart of flesh instead of a heart of stone" signifies newness of life and character and purity of conscience.

Most of these usages are found in the New Testament, but St. Paul has particularly developed some of them. According to him, the heart is the organ of belief as well as of disobedience and impertinence and the immediate receptacle in human life of God's light and knowledge of love, "the eyes of your heart being opened?" These various usages, whether literal or metaphorical, have passed into religious speech, while in the majority of languages, the word "heart" has most of the meanings ascribed to it in Hebrew.

Among the Greeks and Romans, the heart took the place of the liver as the seat of life, soul, intellect, and emotion. Even Aristotle regarded the heart as the center to which all sensory impressions were transmitted. Plato assigned the mortal soul, which governs the intellect and emotion, to the heart, making the brain the seat of the immortal soul, and the liver the seat of the lowest soul and source of sensual desires.

The complex philosophy of the Chinese assigns many *shen*, or souls, to the various parts of the body; that of the heart is supposed to be in the shape of a red bird.

For the Arab–Muslim scholars, the heart had similar physical and moral functions. It is cited to be the center of faith, emotion, etc. It is also the center of all faculties and breath. In Avicenna's *Canon*, he states:

> There is one single breath which accounts for the origin of the others and his breath according to the most important philosophers, arises in the heart, passes them into the principal centers of the body lingering in them long enough to enable them to impart to it their respective

temperamental properties. Lingering in the cerebrum, it receives a temperament whereby it is capable of receiving the faculties of sensation and movement (sensitive faculties), in the liver it receives the faculty of nutrition and growth (vegetative faculties); in the generative glands it acquires a temperament which prepares it for receiving the faculty of generation (reproduction). The foundation or beginning of all these faculties is traceable to the heart, as is agreed upon even by those philosophers who think that the source of visual, auditory, and gustatory power lies in the brain.

(P. 124) (106, 107)

According to Ziskind and Halioua (p. 65) the heart was the receptacle of the vital forces of the universe, maintaining the human body in harmony with the cosmos, which may be disrupted by demons or hostile gods, resulting in diseases.

Rumi, the great Sufi spiritualist and poet, wrote numerous poems and verses about the heart in the 13th century, under the Islamic empire. Amongst his spiritual teachings, in the "the Sufi path of Love" collected by C. Chattick, are the following lines on the heart:

The ultimate center of man's consciousness, his innermost reality, his meaning, as known by God is called the heart. As for the lump of flesh within his breast that is the shadow or outermost skin of the heart. Between this heart and that heart there is infinite levels of consciousness and realizations ...

The Qur'an attributes faith to the heart. The Qur'an mentions that nonbelievers are blind not in their eyes, but it is their heart, which is in the chest, which is blind.

CHAPTER 13

Paleopathology of Cardiovascular Diseases and Pathogenesis with Analysis

S IR MARC ARMAND RUFFER (1859–1917), who was professor of bacteriology at the Cairo School of Medicine at the beginning of the 20th century, was the first to establish Egyptian paleopathology. Preparation of the tissues, which are brittle, was a rather tedious process. The solution giving the best result was: 100 parts alcohol and 60 parts 5% sodium carbonate solution. If the solution softened the tissue too much, more alcohol was added. After a period of time that depended on the bulk and nature of tissue under examination, the solution was replaced by 30% alcohol, and more alcohol was added day by day. After two or three days, the softened tissue was transferred to absolute alcohol, then chloroform and paraffin, and cut using three divisions of Minot's microtome. Suc preparations, after maceration in 1/10,000 caustic potash, gave excellent results. By means of this technique, Ruffer prepared sections of muscles (cardiac, voluntary, and smooth), blood vessels, skin, intestine, stomach, liver, kidney, bone, mammary glands, and testicles. The main characteristics of all these organs and tissues could be readily recognized. The striation of the muscle fibres, the muscularis, the submucosa, occasionally even the glands of the intestines, the convoluted tubules, the straight tubules, the glomeruli of the kidneys, and the various layers of the skin could be identified with certainty, and micro-photographs of them are nicely shown in the book *Studies in the Paleopathology of Egypt* by Sir Marc Armand Ruffer, edited by Roy L. Moodie (1921). In the *Cairo Scientific Journal* (1910, Jan, vol. IV), Ruffer presented his remarks on the pathological anatomy of different structures. Amongst these were the blood vessels in which he demonstrated definite arteriosclerosis (Figure 13.1). We copy, with permission, what he wrote about "Arterial Lesions found in Egyptian Mummies from Dynasties XVIII–XXVII i.e. 1580–527 B.C.". Description of the arteries examined:

1. Aorta (Dyn. XXII) consists of a piece 4½ inches long, covered almost throughout its whole length by a hard calcareous plate.

2. Aorta (Dyn. XVIII–XX) The arch had been hacked away by the embalmer, who had also cut right through all coats just above the bifurcation of the vessel. The thoracic aorta from a point just above the origin of the left subclavian artery and the whole of the abdominal aorta were intact and were easily removed. The internal coat is studded with small calcareous patches, and the two largest, each nearly the size of a shilling, are situated just above the bifurcation. The left subclavian artery at a point just above its origin is almost blocked by a raised ragged calcareous excrescence, as large

PLATE III

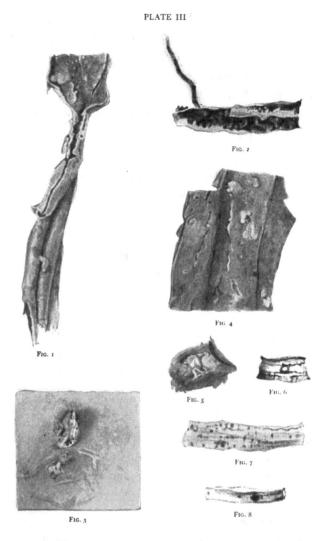

Fig. 1

Fig. 2

Fig. 3

Fig. 4

Fig. 5

Fig. 6

Fig. 7

Fig. 8

FIGURE 13.1   Pelvic and thigh arteries completely calcified (2 Dyn. XX–XXIII). Completely calcified profunda artery after soaking in glycerin. (Dyn. XXI). Partly calcified aorta. (Dyn. XVII). Calcified patches in aorta (Dyn. XVII). Calcified atheromatous ulcer of subclavian artery (Dyn. XVIII–XX). Patch of atheroma in anterior tibial artery (glycerin), the center of the patch is calcified. (Dyn. XXI). Atheroma of brachial artery (glycerin). (Dyn. XXI). Unopened ulnar artery atheromatous patch shining through (glycerin). (Dyn. XXI).

as a threepenny bit (calcified atheromatous ulcer) (Sandison, 1980). Small atheromatous patches, not calcified, are scattered through the whole length of the aorta, and these, owing to the coloration of the tissues, are more easily felt than seen. The common carotid arteries show small patches of atheroma, but the most marked changes are found in the pelvic arteries and in those of the lower limbs.

The common iliac arteries are studded with small patches of atheroma and calcareous degeneration. The other arteries of the pelvis are converted by calcification into rigid "bony" tubes, down to their minute ramifications. So stiff and brittle are they that it was impossible to dissect them out entire, and in spite of every possible care they were invariably broken. The minute intramuscular arteries were easily felt on triturating the muscles under the fingers.

The common femoral and profunda were dissected out. Both were converted into rigid calcareous tubes. It is to be noted that, as far as could be made out from the examination of the cartilages of the ribs, the mummy was not that of a very old person.

3. Atheromatous patches in the aorta and brachial arteries in a Greek mummy. From the examination of the cartilages, etc., I concluded at the time that the man was not above 50 years old at the time of death.

4. Piece of thoracic aorta (Dyn. XXVII) altogether 4½ inches long. It contains seven calcareous patches. No other lesions.

5. Aorta from a Coptic mummy – small hard calcareous patches scattered throughout its length. The two largest are just above the bifurcation and are almost the size of a sixpenny piece (Leriche Syndrome).

6. and 7. Pieces of two aortae, thoracic (Dyn. XXI). No lesions.

From Ruffer's series, it seems that the ancient Egyptians suffered from arterial lesions similar to those found in the present day. Moreover, when we consider that few of the arteries examined were quite healthy, it would appear that such lesions were as frequent 3,000 years ago as they are today.

Regarding the etiology, Ruffer (1921) stated:

> The etiology of this disease 3,000 years ago is as obscure as it is in modern people. Tobacco can certainly be eliminated, as it was not used in ancient Egypt. Syphilis also can be eliminated. Alcohol played a part in Egyptian social life, in so far that on festive occasions some of the ancient Egyptians certainly got drunk as is shown by pictures found in Egyptian tombs. Beer was a common beverage, and wine was not only made in the country, but also imported. However, it is clear that the Egyptians as a race are not and never have been habitual drunkards. From my experience in over 800 postmortem examinations during the Musulman pilgrimages, people who had never touched alcohol in their lives, I have found that disease of the arteries is certainly common and occurs as early in total abstainers as in people who take alcohol regularly.
>
> There is no evidence that the ancient Egyptians worked hard either mentally or physically. Indeed the timetables of workmen, which have been discovered, show that the Egyptian natives of ancient times toiled practically the same hours as the Egyptians do now. They enjoyed a holiday every seven days.
>
> I cannot accuse them of very heavy meat diet. Meat is and has been something of a luxury in Egypt, and although on the tables of offerings of the ancient Egyptians haunches of beef, geese and ducks are prominent, the vegetable offerings are always present in a greater number. The diet

then was, as now, mostly a vegetable one, and often very coarse, as shown by the worn appearance of the teeth. Nevertheless, I cannot exclude a high meat diet as a cause with certainty, as the mummies examined were mostly of priests and priestesses of Deir El-Bahari, who, owing to their high position, undoubtedly lived well.

A good example of overlap between history and medicine are the works of Shattock (1909) and Ruffer on the pathological anatomy of the aorta of King Menephtah, the reputed Pharaoh of the Hebrew Exodus. The mummy was found at Thebes in the tomb of Amenhotep II, who reigned in Egypt from 1499–1420 B.C., and was unwrapped by Dr. G. Elliot Smith, who sent the aorta to the Royal College of Surgeons of London. The finding of Menephtah's mummy at Thebes was questioned by the adherents of the theory that as the Pharaoh of the Hebrew Exodus, he must have been drowned in the Red Sea. Could it be that some cadavers may have washed up onto the shore? Could it be that people found the king's body then?

Elliot Smith (1908) was the one who unwrapped the mummy after confirming its identity, that of Menephtah. He stated:

> All the viscera were removed from the body cavity except possibly the heart. I was able to recognize part of the heart pushed far up into the thorax, but still attached to the aorta … The aorta was affected with sever atheromatous disease, large calcified patches being distinctly visible.

A piece of the aorta, 3 cm in length, was sent to the Royal College of Surgeons in London and was examined by S.G. Shattock (1909). He studied the specimen extensively by various techniques, which can be seen in his article. Shattock stated:

> The section comprises only the middle coat, the presence of inorganic particles of interlamellar substance,

indicating the calcification of the muscle cells and the whole picture accurately reproducing that presented in senile calcification of the media of the aorta.

On a side note: There should be no confusion between Merneptah and Menephtah. Menephtah was the son and successor of Rameses II of Dynasty XVIII and according to the records of the Alexandrian Period; he was the Pharaoh of the Exodus, who was supposed to have drowned in the Red Sea. His body was probably picked up after the Exodus after being thrown on to the shore and was mummified. According to the Qur'an, chapter 10, the Sura of Jonah states that the Pharaoh of the Exodus (Menephtah?) was pursuing the Israelites with his army. When he was on the verge of drowning he confessed to Jahova, the God of the Israelites, that he believed in Him. As a result his body was thrown onto the shore so that his people could take it to be mummified and preserved (see English translation by Mohamad Asad, p. 305).

However, Merneptah was the son and successor of Seti I of the Dynasty XIX, reigning from 1212–1202 B.C. The poetical stela of Merneptah (the Israel Stela in Cairo Museum, No. 34025) recounts his victory over the Libyan army, which invaded Egypt in the fifth year of his reign. A copy of this was found in the temple of Karnak. The poem is long and commemorates his victories over Egypt's neighbors, especially Palestine and Syria. The poem is especially significant for mentioning Israel among the conquered lands and this is the only occurrence of the name of Israel in Egyptian text. A portion of it is here narrated by (Lichtheim, 1976, pp. 73–77):

The princes are prostrate saying "Shalom!"
Not one of the Nine Bows lifts his head:
Tjehenu is vanquished, Khatti at peace,
Canaan is captive with all woe.
Ashkelon is conquered, Gezer seized,
Yanoam made nonexistent;
Israel is wasted, bare of seed,

Khor is become a widow for Egypt
All who roam have been subdued
By the King of Upper and Lower Egypt, *Banere-meramun*,
Son of Re, *Merneptah, Content with Maat*
Given life like Re every day.
Other Paleopathological Studies (Post Ruffer's)

An additional report by Michael R. Zimmerman in 1977, examined 50 mummies from a tomb in Upper Egypt, and he found arteriosclerosis only in one mummy. That shows that there were two classes in Egypt, viz. the wealthy, who had a high-fat diet, and then there were the poor or lower class, who ate mostly vegetables and meat only on occasion. This clarifies the difference between Ruffer's and Zimmerman's respective studies. For more details about nutrition in ancient Egypt, please see the elegant work of Estes (Zimmerman, 1977).

Additionally, Dr. Michael R. Zimmerman, in his excellent work on paleopathology of the cardiovascular system, did experimental mummification by desiccation and rehydration. He studied the tissues histologically and found that the atherosclerotic and thrombosed coronary artery showed excellent preservation of the atherosclerosis and calcification after mummification. More details can be seen in his article entitled "The Paleopathology of the Cardiovascular System" (Zimmerman, 1993).

Long in 1931 reported on the mummy of Lady Teye, from the Dynasty XXI, which showed evidence of atheromatous disease of the aorta, with calcification of the coronary arteries and mitral valve. She also had a nephroarteriosclerosis, which probably caused high blood pressure.

Shaw (1938) discovered arterial sclerosis of the superior mesenteric artery in the mummy of Har-mosĕ, a singer of the Dynasty XVIII.

Cockburn and Cockburn (1975) reported on a mummy from University of Pennsylvania Museum, which showed severe arterial sclerosis of the aorta and diffused arteriolar sclerosis.

Starting in 1972, the mummy program was initiated in Manchester University School of Biological Sciences (1979), studying mummies in detail using a multidisciplinary team including radiology, computed tomography (CT), magnetic resonance imaging (MRI), histological examination (Sandison, 1962), electron microscopy, carbon dating, serological tests, DNA studies, fingerprinting, scientific facial reconstruction, and dental studies. They established a mummy tissue bank. For further reading on the Manchester Mummy Project, please see David (1979).

RADIOLOGICAL STUDIES

Radiological examination of mummies showed calcification of the aorta, femoral, and carotid arteries. Radiological studies have detected calcification of the vessels of Amenhotep and Rameses II, but a radiological survey of Egyptian mummies in European and British museums detected such changes in only four of 88 mummies (Harris et al., 1980; Gray, 1967). This low incidence is probably due to technical difficulties. The article entitled "Egyptian Contributions to Cardiovascular Medicine" by Willerson and Teaf is worth reading (1966).

Ruhli et al, published a new MRI technique (the pulse sequence) to study Egyptian mummy tissues in the Mummy Project of Switzerland without rehydration of the tissue. See the technological details in their letter to JAMA (2007). They state that morphological evaluation of mummified tissue can be conducted without rehydration of the tissues

PATHOGENESIS OF CARDIOVASCULAR DISEASES IN ANCIENT EGYPT

1. Nutritional factors: Higher incidence probably correlates with the social class of the deceased. The higher class had a higher content of fat in their meals, and they were probably more sedentary. On the other hand, the working class rarely had meat in their diet and consumed less fat, and they were

physically active. The average diet contained bread, onions, and a beer almost every day.

The effect of onions and garlic on arteriosclerosis is well documented by the works of Yamagita et al. (2003), Arora and Arora (1981), Vatsala et al. (1980), and Sainani et al. (1979). They demonstrated lower postprandial lipemia after eating a fatty meal with onions and or garlic. In their experiments in rabbits, they found significant effect of onions and garlic in inhibiting the rise in serum cholesterol, serum triglycerides, serum beta lipoproteins, and serum phosopholipids. They had a significant effect in enhancing the fibrinolytic activity. The β:α ratio was altered favorably, being kept close to normal. They found that there was significantly less aortic atherosclerosis in the garlic and onion group than in the pure cholesterol group.

2. Straca.

3. Syphilis in Egypt (Cartwright, 1991; Major, 1954, p. 364–371): According to Moodie, the presence of syphilis in early Egypt is still unproven, and Elliot Smith, Wood Jones, Ruffer, and others deny its existence. On the other hand, Fouquet, Jarricot, Lortet, and others have suggested its presence, and it will be important to look at their evidence. In a large monograph on the mummified animals of Egypt, in the section devoted to anthropology, osseous erosions in the skull of a young woman found at Roda suggested to Lortet the occurrence of syphilis, although he recognized the possibility of the lesions being due to chronic inflammation of uncertain nature, possibly due to tuberculosis. He later supported the idea of these lesions being syphilitic in two contributions in which he defends the idea very strongly. Fouquet's original paper is given by de Morgan, pointing to the prehistoric existence of syphilis, although he has not been

supported in this assumption by subsequent workers. The lesions he found on the prehistoric skull from Amra doubtless may have other explanations; however, a single case is not conclusive.

Jarricot also has suggested the existence of syphilis from a study dating from the Greco-Roman period. Berkhan regards the large size of the head in certain Egyptians as pathologic. However, extensive research in many sources could not find evidence of the existence of syphilis in ancient Egyptian medicine. It was imported to Europe in the 15th century A.D., and its source is not quite definite. The source lies between Columbus's sailors, who came with him back from the Haitian islands; or else from Africa through slaves, who are known to have yaws, which was common in Africa. Also, syphylitic aneurysms have not been reported in spite of extreme vascular paleopathology. Elliot Smith studied 30,000 mummies and Wood Jones 6,000 mummies, but neither of them found evidence of syphilis.

According to Dr. Hassan Kamal (1922), a committee examined the cadavers in the region of the First Aswan Dam and found no trace of syphilis.

4. Diabetes mellitus: There was no definition of diabetes mellitus in the ancient texts, as we know it now. In "Diseases of the Bladder and Urine" Grapow (1955, vol. IV–I, pp. 132–135) writes about excessive micturation (overflow) the writer does not mention other symptoms with this overflow. It may suggest polyuria, in some cases; it may have been due to diabetes. In the recipes of the Ebers Papyrus (Ebers and Stern, 1875; Ebbell, 1937) listed in the following one notices that most of ingredients used are: peeled wheat grains, ochre, gum, juniper fruit, and papyrus grass; honey, raisins, and sweet beer are also added, to be given for four days. One can assume these

recipes are for bladder diseases, probably due to bilharziasis; but it is not clear. In general, there are many names of diseases that cannot be translated or defined from the original Egyptian texts. These prescriptions for frequency of micturition, mostly due to schistosomiasis, existed in Egypt since antiquity. They may have been used for diabetic polyuria also.

Eb. 274 (50, 2–3).

A remedy to remove urine if there is too much of it: peeled wheat grains 1/8; *isd* – fruit 1/8; ochre (*ṣtj*) 1/32; water 5 *ro*; should be exposed to the dew at night, be squeezed out and drunk on 4 days.

Eb. 275 (50, 4).

Another remedy: gum ¼; pealed wheat grains ¼; fresh puree (*h*) ¼; should be squeezed out, should be drunk on 4 days.

Eb. 277 (50, 6–8) = H 63 (4, 14–15).

Another (remedy) for getting rid of too much urine: gum ¼; pealed wheat grains ¼; fresh pap or puree (*h*) ¼; ochre (*stj*) 1/32; water; honey 15 *ro*; should be left overnight, squeezed out, drunk on 4 days.

Eb. 278 (50, 8–9) = H 64 (4, 15–16).

Another (remedy): root of the *k d.t* plant ¼; raisins 1/8; honey ¼; juniper fruits (*w n*) 1/32; sweet beer 7½ *ro*; should be boiled, squeezed out, drunk on one day.

Eb. 279 (50, 10–11) = H 66 (4, 16–5,1).

isd – fruit 1/8; peeled wheat grains 1/8; ochre (*stj*) 1/32; gum 1/32; water 10 *ro*; the same way.

Eb. 280 (50, 11) = H 65 (4, 16).

Another (remedy): gum 1/8; honey 1/32; water 5 *ro*; should be boiled, squeezed out, drunk on one day.

Eb. 276 (50, 5–6) = Eb. 281 (50, 11–13).

Another (remedy) to get rid of urine frequency and escaping *(s). pr.t – snj.* Fruit 1; *gjw* (papyrus grass) 1; beer, one measure of *Hin*; should be boiled, squeezed out, drunk on 4 days.

Eb. 264 (49, 4–6).

Another (remedy) to put right (*k*) abundance of urine. *gjw* (papyrus grass) 1; *pr.t – snj – * fruit 1; root of the *bhh – * plant 1 should be pounded (*hbk*) into a lump, left overnight in sweet beer, drunk together with the sediment at the bottom.

5. Pulmonary tuberculosis seems to have been rare in ancient Egyptians, probably due to warm weather and sleeping outdoors, without crowding. Strong sunrays are also tuberculostatic or bacteriocidal. Ruffer succeeded in finding gram-negative bacteria from the lungs, which he considered to be *Pasteurella pestis* of the Dynasty XVIII. In spite of this fact there are many recipes for the treatment of chest trouble especially cough and pain. Ruffer found Gram-positive bacteria and Shaw demonstrated pneumonia in King Haremheh's lung. (See Grapow, *Diseases of the Internal Organs: Chest Cavity and Lung*, vol. IV–I, pp. 85–86, in German). See "Tuberculosis" in the chapter on pathology.

6. Hypertension: The ancient Egyptians do not mention this disease, but it might have existed based on the following clues:

a. Kidney troubles were detected in Egyptian mummies, e.g. kidney stones, atrophic kidneys, renal bilharziasis, jumbo vesical stones, and nephroarteriosclerosis (Daremberg, 1879).

b. Hematuria was frequently described and even its treatment by (stibium), which is specific for bilharzias (Ruffer, 1921; Smith, 1908; Dawson, 1953; Jonckheere, 1947).

c. Grapow nicely tabulates recipes for treatment of headache and migraine. Some headaches may be due to hypertension (1955).

Another medicine for the head when it is sick, which makes the pain disappear, is prescribed as follows:

Terebinthina resin (Cyprus pine tree), (*sntr*) 1, fat of *bw* – Plant 1 resin (from *Cistus creticus*) 1, reed (*Phragmites communis*), *isw* 1, cattle fat, to be ground (triturated), boiled and used to anoint (*wrh*). Eb. 253 (48, 5–7)

Another medicine for the head when it is sick; this makes pain vanish:

Terebinthina resin (Cyprus pine tree), (*sntr*) 1, Caraway (Carum caris), Juniper fruits (*w n*) 1; goose fat (*mrh.t*) 1, will be boiled and anointed (*wrh*). Eb. 254 (49, 7–9)

Another medicine to treat the head:

tj – sps (cinnamon ?) 1, *ibs* – Plant 1, *gnn* – Part of *hs, j.t* – Balsam 1, Terebinthina resin (*sntr*) 1 to be anointed (*wrh*) everyday.Eb. 255 (48, 9–10)

some of the recipes are translated. One of the interesting recipes called the "holy ones" is the one that Isis made for Re, to expel the disease that was in his head:

Coriander fruit (*s w*) 1, fruit of *h sj – t.* (Bryonia) 1, *S m* – Plant, fruits of *s ms* – Plant 1, *pr. t – snj* – fruit 1, *x* – honey

1 to be made into a lump, will be mixed with this honey, to be bandaged to the head – so that it will make it immediately better.

"When this prescription is done, whichever painful disease, be it a disease of the head – face or head only – be it blood flowing (?) – disease (*h w – snf*), the action is from a god, a goddess, a dead man's spirit or a dead woman's in the head, in any site of the head, so it will be immediately better, really useful". Another remedy:

Berry of the coriander, berry of the poppy – plant, wormwood, berry of the sesames – plant, berry of the juniper – plant. Make into one, mix with honey, and smear herewith. When he uses this remedy against all illness in the head and all sufferings and evils of any sort he will instantly become well.

Sw.t – dhwtj (wild cinquefoil grass) (Potentilla) 1, *nhd.t* (0) 1; caraway 1, juniper fruit (*w n*) 1, *ntjw* – resin 1; resin (*d*) of pine (fir) 1; fat (*d*) of ibex (stone – buck or capra ibex) 1, laudanum (resin from *Cistus creticus*), made into a lump, left at night exposed to dew, then strained to be drunk on four days. Eb. 299 (52, 13–15)

Many of these products are vasodilators. These remedies may have been used to treat hypertension, but there is no way to confirm that they were.

The Ebers Papyrus repeatedly refers to "the bad sickness", by which was meant various forms of diarrhea, dysenteries in different forms, typhoid, and perhaps cholera

7. Collagen diseases (connective tissue diseases) (Ryle, 1948): Spondylitis deformans was not uncommon among the early Egyptians, often of a very severe nature, seen in one vertebral column described by Ruffer and Rietti, belonging to Nefermaat, a man in Dynasty III (2980–2900 B.C.); the

whole vertebral column was ankylosed. Less severe forms and mild lesions were also detected. This may account for collagen diseases present in ancient Egyptians. But collagen arteritis most probably did not exist at that time since arterial examination of mummies of young individuals did not show sclerotic lesions.

CHAPTER 14

Ancient Egyptian Pharmacology

ETYMOLOGY

Before proceeding with this chapter, a little etymology might be interesting. Pharmacology is that science that deals with drugs: their sciences, their characteristics, their chemistry, and their uses. Under pharmacology, there are several subheadings, e.g. pharmadynamics, pharmakinetics, pharmacotherapy, pharmacognosy, etc. The etymology of the word *pharmakon* is from Greek *ph-ar-maki*, which means "that which brings peace". This is an expression found on an Egyptian relief of god Thot on the head of Ibis.

The word "alchemy" came from an Egyptian word *kham* (*chim* in Greek), which means black, probably because the mixtures in the laboratory retorts looked black, since the Egyptians prepared a lot of chemicals for painting and other industrial purposes. The word was modified by the Arabs to *alkimiaa*, which was introduced to Europe as alchemy; in German it is *Alchimie.*

Horus eye, the eye of the healing god Horus, was used as an emblem for practitioners, and it was thought to be the origin of the prescription sign, Rx, which is used worldwide.

The caduceus, which is the snake symbol, was an important emblem in medical practice both in Egypt and Babylonia at approximately 4000 B.C., and the Greeks attributed it to Hermes, who is reciprocal to Thoth of Egypt. Thoth was at the top of the Egyptian healing gods. He was represented as an ibis, but later was represented as a man with the head of an ibis, was surmounted by the solar disc and lunar crescent. He was known to be a great scientist, known as the measurer and mathematician, and developed accurate sciences of art, theology, occult sciences, magic, and secret medical formulas. According to Clemens Alexandrius of the Christian era, Thot (Hermes of the Greeks) compiled 42 volumes in different fields. He was identified later by the Greeks as Hermes Trimegistus. Georg Ebers believes that his papyrus is one of these volumes. The caduceus is still used as a symbol in the medical corps and other associations. In Greek caduceus is the wand of Hermes, and it is made of a short herald staff with two serpents entwining around it in a double helix and sometimes surmounted by two wings. In antiquity, it was used as an astrological symbol representing the planet Mercury. It should not be confused with the traditional medical symbol, which has a short rod with only one serpent.

The Egyptian doctor used to prescribe the remedy in a way not different from ours nowadays. The Egyptian pharmacist used to supply infusions, decoctions, macerations, fumigations, inhalations, gargles, injections (vaginal or rectal), pills, powders, triturations, salves, suppositories, confections, cataplasms, poultices, etc. Many of them are still used, and some of them have a sound scientific basis (Helmuth, 2000), e.g. ammi visnaga, milanidin, castor oil, cosmetics, etc. It is beyond this monograph to describe all the remedies (cf. Dragendorff, 1898). Unfortunately, the names of many plants could not be deciphered, and perhaps many of them have kperished totally. The efforts of Dr. Grapow and his collaborators in this field are immense. Also, Dr. Vivi Täckholm of Cairo University made a great effort to study the flora of ancient Egypt, concurrently with investigation of the ancient names. Apart from Ghalioungui's publications, there are two recent texts on flora that are reviewed, namely, Moursi and Manniche.

In 1939 an Egyptian pharmacist compiled a book on the history of pharmacy and chemistry in ancient Egypt. He translated some of the flora of Egypt into Arabic (Abdul Rahman, 1939). We will review some of the medicinal plants used in cardiovascular diseases, both locally and generally. They have been abstracted from the recipes specifically chosen by Grapow et al. (1935) in their chapter "Herz und Geffässe" in vol. IV–I of *Grundriss der Medizin der Alten Aegypten*. For these medicinal plants referenced, we will be listing their principal actions based on the experience of the ancients which was often confirmed by modern science. The main source used here is Ebers Papyrus (1875), translated in 1937 by Ebbell, but he was not certain of some of the prescriptions, which total over 842. From this text, we are selecting prescriptions that have possible actions in cardiovascular disease. Ghalioungui's text about Ebers Papyrus is relatively more recent (1987) but is not commercially available.

These botanical and herbal sources are enumerated below and their pharmacology is briefly presented for general knowledge: acacia, anise, barley, caraway, cassia and cinnamon, castor plant, cedar, colchicum, colocynth, coriander, corn seeds, cumin, date blossoms, elderberries, emmer wheat, fennel, *Ficus sycomorus*, figs, fir oil, frankincense, garlic, grapes and raisins, juniper, king's clover, linseed, henbane, lotus flowers, mandragora, mastic, myrrh, onions, peppermint, poppy, saffron, sesame grains, squill, tamarinds, and turpentine.

Compare also Leak (1940), Sallmann (1957), Schering Corporation (1975), Trease (1966), Wallis (1967), and Wren (1994). Now some of the uses will be discussed:

Acacia Senegal (Acacia Gum)

Demulcent, mucilaginous, often used as an ingredient in medicinal compounds for the treatment of diarrhea, dysentery, catarrh, cough, hoarseness, fevers, etc.

Acacia Nilotica del. (Acacia Arabica Willd.)

The pod is known as "sant" and the bark as "babool". Sant has long been used for tanning and as a medicinal astringent. The tannin is of the catechol type. The deseeded pod contains 30–45% of tannin. The Masai (in Africa) use a decoction of the bark as a nerve stimulant. In Africa and Egypt, the gum, bark, and leaf are used medicinally in colds, ophthalmia, diarrhea, and hemorrhage. The leaf was used as an emmenagogue, the flower as an ointment and the fruit as a remedy for diarrhea and gynecological conditions. The bark and fruit are astringent. In Ethiopia the wood is used as a cure for smallpox. The gum is edible and contains: 14.49% moisture, 2.41% ash containing 0.765% Ca and 0.106% Mg, 50.43% pentosan, 21.85% galacton and yields 1–arabinose and d–galactose by hydrolysis

> The seed yields 5.46% of yellow fixed oil. The green parts yield 200 mg of vitamin C and 4800 i.u. of carotene.

> A decoction of the bark and root is drunk to act as tonic and aphrodisiac. The contused young leaf is used as a dressing for ulcers.

In pharmacology tannins in general are antiseptic against bacteria and influenza virus, constipative, and astringent.

Ficus Sycomorus (Sycamore Fig)

The fruit is eaten in Egypt. The fruit and leaf are fed to cows to increase the flow of milk. A decoction of the bark and its latex is used for chest conditions, cough, and scrofula. The juice is used for inflammations; the bark is antidiarrheal. An infusion of the bark relieves throat pain. The plant contains ferment (chymase). The leaf, stem, fruit, and root give positive tests for flavonoids. It is a folk remedy of frequent use in Egypt, but its chemistry has not been fully studied.

Ficus Carica (Common Fig)

Figs were a valued article of food and medicine amongst the ancient Egyptians and subsequently the Hebrews and the Greeks. In the Qur'an (Rodwell, 2004) it is mentioned in relation to Sinai. The Hebrews called it *tinab*; in Arabic it is *teen*. It was one of the principal fruits of Palestine even before the entrance of the Hebrews. The fruits were made into a cake for the treatment of boils. The fig tree was associated with the wine as an emblem of peace or prosperity. Failure of the crop and destruction of the tree were regarded as misfortune and punishment from God. The tri-lobed leaf remained in symbolic worship since the earliest date. Let us read a quotation from *Cultus Arborum*: "Near Cairo at a fountain wherein the Virgin Mary washed her infant's clothes, a lamp was, three centuries ago, kept burning in her honor in the hollow of an old fig tree, which has served them as a place of shelter" according to the *Intinerario de Antonio Tenreiro* (Lloyd, 1929). The fresh latex of *Ficus* (fig tree) species is antihelminthic. Its active principal is ficin, a crystallizable protective enzyme, which digests *Ascaris* in vitro. Ficin is highly effective in preventing the coagulation of blood and of milk, by digesting prothrombase and caseinogen (Cançado, 1944).

Sesamum Indicum (Sesame Seeds)

Gingelly seed, teel seed, benne seed. Constituents: The seed contains fixed oils 45–55%, proteins 15–20%, mucilage 4%. The oil consists of glycerides of oleic, linoleic, palmitic, stearic, and myristic acids, also a crystalline substance, sesamum, and a phenolic substance, sesamol. The leaves are used as an aphrodisiac decoction, for malaria and for cough and for sores (tannin and chlorazemic acid). The seed and its oil are used as emmenagogue, abortifacient, and purgative. The oil is used as a base for fish canning, cosmetics, and margarine. The seed cake, after expression of the oil, is toxic to livestock, causing colic, tremors, dyspnea, cough, and depression. Cattle fed on excessive sesame meal develop eczema, with itching and hair loss. The oil is used to prepare injections of antibiotics, hormones, and other oil-soluble principals. It contains no vitamin A.

Melilotus Alba (Clover)

When damaged may cause poisoning in livestock, the first sign being delay in the coagulation time of the blood. Severely affected animals die from hemorrhages all over the body. The toxic symptoms develop after at least three weeks of feeding. It is due to the presence in spoiled clover of dicoumarol, which lowers the prothrombin concentration of the blood. In human poisoning, the result is hemorrhage in different parts of the body. All types of *Melilotus* contain a volatile odoriferous principal, coumarin, which in excess leads to fatal paralysis. Sweet clover is not toxic, nor is the dried plant. The seed yields 6.63% fixed oil. Extracts from the leaves, stems, and roots have an action on *Mycobacterium tuberculosis*. The root yields a saponin of type II that causes hemolysis at pH 5.6.

Juniperus (Juniper)

Juniper contains cedrol (cedar camphor) in its wood; it is abortifacient. The leaves produce contact dermatitis. From the leaf, a cold aqueous extract is tuberculostatic, but it has no tumor-damaging property. The seed of some species of *Pinus* is edible, while other species yield turpentine and resin. *Pinus halepensis* is used in the Mediterranean area as folk medicine.

Fructus Carui (Caraway Seeds)

This yields by distillation from 3.5% to 7% of volatile oil, the principal constituent of which is carvone (50–60%). It also contains about 20% proteins and fixed oil in the endosperm and yields about 6% of ash. Caraway or its oil is used extensively as a carminative, a stimulant, against flatulence, and against stomach derangements.

Cuminum Cyminum (Cumin Fruit Seeds)

Yielding 3–4% of volatile oils, whose chief constituent is cuminic aldehyde. It is used as a carminative and stimulant.

Citrullus Colocynthis (Bitter Apple)

The fruit contains a bitter substance. It contains an alkaloid, which produces very drastic purgation, and resin, which is soluble in ether and chloroform and is also a powerful purgative. It also contains an alcohol, citrullel, and a glucoside of cucurbactin E (and elaterin); neither of these are purgative. It is used as a cathartic in dropsy. In India the kernel is used as a diuretic and tonic. The diuretic action is due to the presence of arginine and citrulline. Citrullin was found to triple the output of urea, thus reducing NH_3 level in the blood in liver failure. The seed shell contains 0.06% mannitol. The juice of the root is said to be hemostatic in hemorrhage after abortion. Curcurbicitrin is a glucosidal saponin, which is used in the treatment of hypertenstion, marketed as citrin capsules. Tapeworm and roundworm are paralyzed by an extract of the husk and the kernel.

Ricinus Oil Plant (Ricinus Communis, Linn.)

The leaf is applied to the head to relieve headache, as a poultice for boils; for dressing wounds and sores, mashed leaves are used. It is also for rheumatic pains and applied to the breast as a galactogogue. The root is used with the leaf for rheumatism and sciatica. The dried root is a febrifuge and is used for the treatment of jaundice. The unbroken seed is used as a purgative. One pounded seed produces strong purgation; two seeds may produce fatal sanguinous diarrhea and vomiting. To a child, one seed is fatal. Acute poisoning with seeds results in hyperplastic mycosis, followed by hemolytic anemia, neutropenia, and eosinophilia.

Ricinus oil is produced from the seeds. It contains the toxic albumin ricin and toxic crystalline nitrogenous body, ricinine, and the pressed seeds yield 0.3% ricinine. The latter is also extracted from the leaves. Sodium hypochlorite causes the toxicity of ricin to disappear. Commercial processing destroys a considerable portion of the toxic and allergenic activity of the seed. It contains also riboflavin, nicotinic acid, lipase, uric acid, urease, glutamine, and euglobulin. The entire seed is very actively poisonous due to the ricin, which is not present in the oil but in the cake. Castor oil is bland; when a portion of the unsaturated fatty acid is swallowed, ricinoleic acid is freed as the sodium salt, which produces irritation and purgation. It also contains tocopherol.

Bryonia (*Bryonia Dioica*, Jacq. *Bryonia Alba*, Linn.)

Called also madragora, Lady's seal. The part used is the root. It is irritating, hydrogogic, cathartic, useful in small doses for cough, influenza, bronchitis, and pneumonia. It is valuable in cardiac disorders caused by rheumatism or gout, also in malaria and zymotic diseases. Large doses are dangerous. It is used as liquid extract and as bryonin.

Cinnamon (*Cinnamomum zeylanicum*, Nees.)

The bark contains volatile oil, resin, gummy extract with tannin, lignin, bassorin, sugar starch, fixed oil, and coloring material. The bark is used as an aromatic, astringent, stimulant, and carminative. It is rubificient in headache and a remedy for tuberculosis. It contains volitale oil, which is rich in cinnamic aldehyde. Emgenol is found in the oil at 10%. The oil is germicide, fungicide, and antihelminthic. Swallowed in excess, it is an irritant and may be fatal. The bark also yields 6.6% oxalate, tannin, and mucilage. Taken after meals, it may reduce postprandial glycemia.

Coriander (*Coriandrum Sativum*)

Coriander fruits contains up to 1% volatile oil (oleum corandri B.P & U.S.P.). This contains 65–70% coriandrol (d–linalool and pinene). It is stimulant and carminative, prevents griping, and is flavoring.

Tamarinds (*Tamarindus Indica*, Linn.)

The fruit is edible alone or with other types of food. With water it forms a refrigerating drink, especially in fevers. It is good for digestion and is laxative. It contains oxalic, tartaric, citric, malic, acetic, and succinic acids, and sugars and pectin. The fresh leaf is chewed, and in Tanzania the pulp is packed into the wound caused by a snakebite. The powder is used for wound dressing. The seed contains starch, albuminoids, fixed oil, reducing sugar, and mucilaginous material. An infusion of the leaf is vermifuge. The boiled leaf is antirheumatic. The decoction of the bark is anti-histaminic and emmenagogue. In northern Nigeria, the root is a component in anti-leprosy treatments (Watt and Gerdina, 1962).

Before describing the effect of onions and garlic, it is interesting to quote from Herodotus (1996) (Book II, Chapter 125):

> There is an inscription inside the pyramid, which is written in Egyptian character. It tells us of the quantity of radishes, onions, and garlic that were consumed by workers building the pyramids. I remember most exactly that the interpreter who deciphered the inscription for me remarked that the sum of money spent on those items would amount to 1,600 talents of silver. (This is about $415,626.87 spent for 20 years.)

From the antibiotic effects of these plants, epidemics could be prevented or at least minimized. Compare this with the epidemics during the digging of the Panama Canal.

Other sources of antibiotic action included: the moldy wheat loaf, spoilt meat, yeast, rotted stems of lotus flowers and other plants, soil (for fungi) from different sources, etc. Scientists in modern times have verified the value of these substances. See human and animal experimentation, which confirms the efficacy of ancient Egyptian pharmacopeia, by J.W. Estes (1989). He concludes, "As a result we are not very surprised that those ancient remedies from the banks of the Nile have occasionally resurfaced, even in the modern times."

Onions (*Allium Cepa* L.)

The actions of the onion have had great medicinal popularity through the ages. It is mentioned in the Holy Books. The actions are:

1. Lowering blood sugar, even in the pancreatectomized animal, with maximum effect after one-to-five hours. An alkaloid was isolated; fresh onion juice is also hypoglycemic. Subcutaneous extract injection has the same effect (Watt and Gerdina, 1962).

2. An extract slows the frog's heart and depresses its muscle. But Kreitmair, experimenting with the freshly pressed juice and a solution of it, found that it contains a heart stimulant, which increases its rate, increases systolic pressure, coronary flow, and pulse volume. The heart-stimulating fraction is alcohol soluble (Kreitmair, 1956).

3. Increases the fibrinolytic activity of the blood when added to it. It also prevents reduction of this activity with fatty meals. The active component is not yet known (Estes, pp. 66–71).

4. On the smooth muscles, it has a stimulant action: it stimulates the gut, uterus, and gall bladder, thus promoting biliary flow.

5. It is a diuretic (Kreitmair, 1956).

6. It is bactericidal due to thiocyanic acid.

   Orally, the extract gives good results in the treatment of influenza, gastritis, chronic colitis, and whooping cough. But it has no effect on *Pseudomonas* spp. Pressed onion is more effective than segmented onion in culture media. Eight days' cool storage and boiling for ten minutes greatly reduces its antibacterial power. In the intact bulb, there is no pyruvic acid, but it is formed from precursors when the tissue of the onion is wounded. The outer-pigmented scales have a dyeing property, due to the presence of a coloring material, whose crystals have the formula $C_{15}H_{10}O_7$ and is called quercetic, which on decomposition gives phloroglucin and protocatechinic acid. It also contains quercetin-mono-d-glucoside. The resistance of the onion to fungal diseases has been ascribed to its pyrocatechinic acid content.

A volatile extract gave better results in throat infection than penicillin lozenges. Inhaled fumes were not effective against tuberculosis.

In Russia, it was proved that the fresh volatile substance is highly effective in the treatment of infected wounds, both in animal experiments and in man. It was found to be deleterious to infusorium *Paramaecium caudatum*, upon the growth of *Brucella abortus*, and *Staphylococcus aureus*. Highly colored onions stored for six months are as active as freshly harvested ones. They are inactivated by boiling for five minutes, but not when heated to 60°C. From onions, a volatile oil is isolated, which kills *Staph. aureus* and protozoa in a dilution of 1:100,000. The brown scaly leaf yields a small amount of tannin, quercetin, and other substances. The oil contains allylpropyl-disulfide, a nitrogen-free distillate (lacrymatory), thio-aldehyde, thio-popaldehyde, or thio-acetyldehyde. Hydrogen sulfide is liberated when it comes in contact with phenylhydrazine.

7. It has a strong gonadotrophic effect.

8. After absorption, the oils are excreted through the skin and the lungs.

9. The oils are antispasmodic and anithelminthic.

Garlic (*Allium Sativum* L.)

This ancient plant was used in many folk remedies; recently, it has proved to be very valuable. It has some actions similar to those of onions, viz. it is:

1. carminative

2. antiseptic

3. diuretic

4. hypotensive

5. antihelminthic

6. diaphoretic

7. stimulant

8. expectorant

9. aphrodisiac

10. antiscorbutic

11. antirheumatic

12. rubificient

13. febrifuge

14. stomachic

15. repellent to snakes and mosquitoes

16. stimulative of wound healing

But it is dangerous if excessively applied to children and may even be fatal.

Recently it was found that:

1. Sativine (extracted from garlic) promotes epithelial and granulation tissue growth.

2. The volatile substance is bactericidal, both in animal experiments and in man: this substance contains polyvinylsulfides, aromatic lactones, and probably phytosterols.

3. Alliin is the basic form from which diallylsulfide is formed by enzymic cleavage when the clove is contused. Allicin, derived from alliin, has the formula $C_6H_{10}OS2$. Dilutions of allicin of 1:125,000–1:85,000 are antibiotic against both gram-negative and gram-positive organisms, e.g. staphylococci, streptococci, *B. typhosa*, *B dysenteriae*, *B. enteritidis*, and *V. cholerae*, and to acid-fast bacteria. One microgram of allicin corresponds to 15 Oxford units of penicillin.

4. Cavallito and Bailey isolated allicin in 1944 (Cavallito and Bailey, 1944). As it is unstable, they used a more stable diethyl analog for its effect on malignant cells in animals. The results show that there is an inhibitory effect when this substance is placed in direct contact with the malignant cells, whereas the results are inconclusive when the contact is not direct. Immediately the plant is damaged, the enzyme alliinase splits alliin, with the formation of allicin, which has the odor of garlic and is antibacterial. Further decomposition of alliin yields the sharply odorous allylsulfides. Tests have shown that synthetic alliin in doses of 100–200mg/kg does not retard the growth of various experimental tumors in the mouse (Mutter, 1965); however, this is not confirmed by all scientists. Hsing et al. (2002), found in epidemiological and laboratory studies that allium vegetables (onions, garlic, scallions, leeks, shallots, and chives) have antitumor effects. In their epidemiologic studies, they found that allium vegetables significantly lowered the incidence of prostatic cancer. They are rich in flavonols and organosulfur compounds, which are antitumorgenic in the laboratory. More details can be seen in their paper. A review article by Fleischauer and Arab (2001) stated, "The strongest claim for a protective effect of raw/cooked garlic consumption can be made for stomach and colorectal cancer".

5. An ether-soluble, steam-volatile alkaloid substance has been extracted from garlic. This substance injected into dog and rabbit has a hypoglycemic action, although it is not considered analogous to insulin. That is why garlic is used as a remedy for diabetes in India.

6. Intramuscular administration of alcohol extracts of garlic over a four-day period quadrupled the excretion of 17–ketosteroid in an animal on a lucerne diet. This indicates a corticotrophic effect on the adrenal cortex.

7. Dialysis of the whole clove with ether for six hours gives strong antibiotic, which is active against the colon group of bacteria. It is absorbable and is distributed in the body tissues, including the C.S.F. It is eliminated in bile and urine and still maintains its antibiotic activity. It is an amebicide and may be of value against protozoa, especially those invading the liver.

8. Garlic extracts are also fungicidal in high concentrations and fungistatic in lower concentration. They inhibit *Candida albicans*, *Trichophyton cerebriform*, and *Trichophyton granulosum,* and have the most marked effect on the typhoid–paratyphoid group.

9. Garlic is antihelminthic; it produces strong stimulation of *Ascaris lumbricoides* followed by paralysis due to the diallyldisulfide content. It acts also against oxyurius. Mashed garlic 1–3 g/kg in the food of geese is 80% effective in removing cestodes.

10. The chloroform-soluble fraction is a weak hypotensive and has a slight stimulant effect on the frog's heart. It inhibits the movements of the isolated intestine.

11. The chloroform-insoluble fraction produces a marked prolonged drop in the blood pressure of the cat and a very slight effect on the frog's heart and very slight stimulation of the intestine.

12. Alcohol extract increases the tone of the uterine muscle; it also increases the frequency and amplitude of uterine contractions. *In vitro*, it counteracts the action of procaine on the uterus.

13. The tincture leads to:

    a. slight initial vasoconstriction.

    b. increased coronary flow due to vasodilatation and lowered tone of the heart muscle.

    c. slowing the frog's heartbeat down, due to vagal stimulation and direct action on the muscle.

    d. stimulation of the accelerator mechanism of the atropinized heart.

14. The oil absorbed from the intestine is excreted by the lungs. It leads to formation of methemoglobin.

15. Essence of garlic has a stimulant-narcotic action in all animals.

16. Garlic preparations have been found to prolong the life of mice that have received a lethal dose of Vigantol, an ergosterol-like preparation. They are supposed to be antisclerosing, and on the continent, especially in Germany, they are used extensively in geriatrics.

Details are seen in appropriate literature, especially by Stoll and Seebeck (1951), where the combined antisclerotic effects of onions and garlic have been discussed.

Celery (*Apium Graveolens* Linn.)

The seeds are used in medicine as antirheumatics, emmenagogues, abortifacients, and diuretics (due to glycolic acid). The fruit is poisonous in overdose. The root and leaf are diuretics. The herb contains the glucoside apiin, which is identical with that isolated from parsley. It yields a volatile oil (also gained from the seed) with characteristic smell. The camphor of the oil is known as apiol, which stimulates the gravid and non-gravid uterus. The herb contains carotenoid pigments and thiamin. It is carminative and tonic.

Now we will discuss some plants according to Egyptian sources (Kamal, 1967).

Lotus

This is a classical name for various plants. These include the jujube tree and the sacred water lilies, *Nymphaea lotus*, of Egypt. It was held sacred because the Egyptians saw it as a symbol of the rising again of the Sun. As an amulet, it signified the divine gift of the eternal youth. The root is starchy and was taken as a sedative, soporific, and antispasmodic. To it was ascribed the property of producing sterility. The following are examples of its uses:

1. Anti-epileptic – Ebers Papyrus prescription 209 (Ebbell translation, p. 55): to treat an obstacle on the right side, when epilepsy has passed through him – in a mixture by mouth.

2. Against hematuria (most probably bilharziasis) – as a sedative: Ebers Papyrus prescription 224 (= Hearst Papyrus prescription 82) leaves of lotus – in a mixture by mouth (Ebbell translation, p. 56).

3. Against jaundice – Ebers Papyrus prescription 479: in a mixture by mouth (cf. Ebbell translation, p. 80) (Major, p. 102).

Ricinus

Ebers Papyrus prescription 251. (Translation by Ebbell, pp. 59–60).

To know what is made with the ricinus plant according to that which was found in the old writings as something useful to men: "if its roots are crushed in water and applied to a head, which is ill, then he will get well immediately, like one who is not ill. But if a little of its seeds is chewed with beer by man with looseness in his excrements, and then it expels the disease in the belly of the man. Furthermore, for the hair of a woman it is made by means of its seeds: it is ground mixed together and put into oil by the woman, who shall rub her hair therewith. Additionally, the oil in its seed is used to anoint one who (suffers) from the rose (?), with bad putrid (*ittt*), then (*rjwmw*) the skin (?) as (in) one whom nothing has befallen. But, rubbing the aforesaid for ten days treats him, rubbing very early in the morning until it is expelled. Really excellent (proved) many times!"

Onions (from the Papyri)

Ebers Papyrus prescription 30, Ebbell, p. 67	in a mixture per os against asthma
Ebers Papyrus prescription 828, Ebbell, p. 112	as a vaginal douche in metrorrhagia
Ebers Papyrus prescription 634, Ebbell, p. 94	in a linament to soften the knee
Ebers Papyrus prescription 657, Ebbell, p. 96	to soften the limb
Ebers Papyrus prescription 660, Ebbell, p. 97	in dressing against priapism
Ebers Papyrus prescription 884, Ebbell, p. 113	to prevent a serpent from coming out of its hole
Kahum Papyrus prescription 28	to know if a woman will bear a child or not. (cf. Grapow et al., in *Grundriss der Mediz. der Alt. Aegypten*, vol. IV–I, p. 273).

Christ's Thorn (*Ziziphus Spina Christi*)

It was prescribed for the following conditions:

1. To stop hematuria – taken by mouth.

   Ebers Papyrus prescription 226, Ebbell, pp. 56–57.

2. To fasten a broken bone to unite: local treatment.

   Hearst Papyrus prescription 221, cf. Old Med. Pap. by Chauncey Leake, p. 94.

3. To put the urine in order – local to the penis.

> Ebers Papyrus prescription 272, Ebbell, p. 61.

4. Against abdominal distention.

> Berlin Papyri prescription 159, a rectal enema, Hermann Grapow et al., *Gundriss der Med. der Alt. Aegypten*, vol. IV–I, p. 107.

5. Against impotence in an ointment.

> Ebers Papyrus prescription 663, Ebbell, p. 97.

6. Against indigestion in a local mixture.

> Ebers Papyrus prescription 208, Ebbell, p. 54. cf. Grapow et al., in Drogennamen, pp. 300–302.

There is extensive *materia medica*, which cannot be adequately summarized in this short text. Since Ebers Papyrus is the oldest and most educational text on this subject, it is used as the main source of pharmacological knowledge in general and with special emphasis on cardiovascular diseases.

The sources of the *materia medica* are plants, animals, minerals.

Here we will focus on the plant sources, which are the most common and most amenable to research.

The glossary of the drugs used in pharaonic Egypt has been very well tabulated by Estes. It includes plant, animal, and mineral components and includes some specific and nonspecific indications, and it is alphabetically tabulated. The uses are mentioned briefly, and references are recorded separately. About 350 titles are presented and include gastrointestinal ailments, including parasites, anal complaints, respiratory disorders, internal organs, female diseases, skin diseases, eye complaints, lack of appetite, pains, and insomnia. Regarding cardiovascular diseases, no subject indications were mentioned. The tables include not only Ebers papyrus but also other papyri.

We will review the *materia medica* from Ebers papyrus and those that have been studied in modern times. To avoid repetition, here are just a few that were studiously gathered by Moursi. Those botanical drugs studied in modern times will be presented. They are presented here in alphabetical order by their Latin name. Please note most herbs used for medicinal purposes are not approved by the United States Food and Drug Administration (FDA).

1. *Ammi visnaga*

    Synonyms: Bishop's weed, khella, greater ammi

    Active ingredients: Khellin, visnagin, khelloil, khellol glucoside

    Medicinal parts: Leaves

    Actions: Intensifies coronary and myocardial circulation, acting as a mild inotrope. It acts as an antispasmodic for the smooth muscles.

    Uses: Angina pectoris, cardiac insufficiency, paroxysmal tachycardia, extrasystoles, hypertension

2. *Ceratonia siliqua*

    Synonyms: Carob, St-john's bread, John's bread

    Active ingredients: The main constituents of carob are large carbohydrates (sugars) and tannins.

    Medicinal parts: The fruit and bark, carob seed flour (ground endosperm of the seeds)

    Actions: Hypoglycemic and hypolipidemic effect due to the increase of gastrointestinal tract enzymes and hormones

    Uses: It is antiexudative (for infantile diarrhea). It also has antiviral and anticoagulant effects. In hypocholestermia it is used (seed flour) as a part of low-cholesterol, low-fat diet.

3. *Citrullus colocynthis*

Synonyms: Colocynth, bitter apple, wild gourd (biblical), gall (biblical)

Active ingredients: Ether–chloroform-soluble resin, a phytosterol glycoside (citrullol), and other albuminoids. Bitter substance is colocynthin and colocynthetin.

Medicinal parts: The dried pulp

Actions: Alterative, diuretic, antipyretic, hemostatic

Uses: It is a powerful drastic hydragogue cathartic that produces, when given in large doses, violent griping with, sometimes, bloody discharges and dangerous inflammation of the bowels. In India, it is used for ascites and elephantiasis that are possibly of a vascular nature. It is mentioned in Ebers Papyrus as an ingredient to abort a pregnancy:

To cause a woman to stop [terminate] pregnancy in the first, second or third period [trimester]: unripe fruit of acacia; colocynth; dates; triturate with 6/7th pint of honey. Moisten a pessary of plant fiber [with the mixture] and place in the vagina.

4. *Coriandrum sativum*

Synonyms: Coriander, cilantro

Active ingredients: Volatile oil, fatty oil: chief fatty acids – petroselinic acid, oleic acid, linolenic acid; hydroxycoumarins: including umbelliferone, scopoletine

Medicinal parts: Seeds, leaves, oil

Actions: Antihalitosis, carminative, expectorant, narcotic, stimulant, stomachic

Uses: It stimulates gastric secretion. It is used for dyspepsia, indigestion, loss of appetite. In animal studies it is a strong hypocholesterolemic agent.*

5. *Cucumis melo*

Synonyms: Melon, muskmelon, cantaloupe, honeydew, sugar melon

Active ingredient: Melonemetin

Medicinal parts: Seeds, fruits

Actions: Antioxidant, anti-inflammatory, purgative, emetic

Uses: Apart from being part of the Egyptian diet, it was also a kind of medicine. A remedy to treat the heart (Ebers 220):

Šspt – melon 1/32, notched sycamore fruits 5 *ro*, ochre 1/32, fresh dates 5 *ro*, honey 5 *ro*, water 20 *ro*; is left in the dew overnight, strained, and then drunk for one day. (1 *ro* = 15 ml)

6. *Ficus carica*

Synonyms: Fig

Active ingredients: Amino acids, vitamins, enzymes, linoleic acid

Medicinal parts: Seeds, fruits

Actions: Anti-cancer; demulcent; digestive; emollient; galactogogue; laxative; pectoral

Uses: Figs were used by ancient Egyptians for constipation, abdominal diseases. (see Ebers' prescriptions number 202, 92)

---

* Chithra, V. and Leelama, S. Hypolipidemic effect of coriander seeds (*Coriandrum sativum*): mechanism of action. *Plant Foods Hum Nutr.* 1997; 51(2):167–172.

7. *Melilotus officinalis*

Synonyms: Sweet clover, ribbed melilot, field melilot, cornilla real

Active ingredients: Flavonoids, coumarins, tannins, volatile oil

Medicinal parts: Flowers, leaves, root, seedpod

Actions: Alterative, diuretic, antipyretic, hemostatic

Uses: Also anti-exudative and anti-edema. It increases venous reflex and lymphatic outflow. It is used for chronic venous insufficiency and post-phlebitic syndrome.

8. *Morus alba* and *Morus nigra*

Synonyms: White mulberry and black mulberry (presently the former is used more frequently, but they can be used interchangeably as they have the same properties)

Active ingredients: Fruit: glucose, protein, pectin, coloring matter: tartaric and malic acids

Medicinal parts: Fruit, inner bark, leaves, manna

Actions: Analgesic, anthelmintic, antibacterial, antirheumatic, astringent, diuretic, expectorant, hypoglycemic, hypotensive, purgative, sedative

Uses: Recent research has shown improvements in elephantiasis when treated with leaf extract injections and in tetanus following oral doses of the sap mixed with sugar.* Experimentally, according to Xia et al., the leaf extract produces aortic vasoconstriction followed by vasodilation.† The stems are antirheumatic, antispasmodic,

---

* Bown, D. *Encyclopaedia of Herbs and their Uses*. Dorling Kindersley, London. 1995.

† Xia, M.L., Gao, Q., Zhou, X.M., Qian, L.B., Shen, Z.H., Jiang, H.D., and Xia, Q. Vascular effect of extract from mulberry leaves and underlying mechanism. [article in Chinese]. Zhejiang Da Xue Xue Bao Yi Xue Ban. 2007 Jan; 36(1): 48–53.

hypotensive; they are used in the treatment of rheumatic pains and spasms, especially of the upper half of the body, high blood pressure.

9. *Myristica fragrans*

Synonyms: Mace, magic, muscadier, muskatbaum, myristica

Active ingredients: Myristicin, safrole, and methyleugenol are key components.

Medicinal parts: Seed

Actions: Antibacterial, analgesic, diaphoretic, febrifuge, stomachic

Uses: In animal experiments by Sharma et al., it lowers cholesterol and triglycerides, lowers the cholesterol to phospholipids ratio, as well as elevated HDL ratio. Feeding on the seed extract prevents accumulation of cholesterol, triglycerides, and phospholipids in the liver, heart, and aorta. It also dissolved athermanous plaques in the aorta. Fecal excretion of cholesterol and phospholipids increased significantly.*

10. *Punica granatum*

Synonyms: Pomegranate

Active ingredients: Alkaloids: isopelletierine

Medicinal parts: Fruit, leaves, seed

Actions: Antibacterial, antiviral, astringent, demulcent, emmenagogue, stomachic, vermifuge

---

* Sharma, A., Mathur, R., and Dixit, V.P. Prevention of hypocholesterolemia and atherosclerosis in rabbits after supplementation of Myristica fragrans extract. *Indian J Physiol Pharmacol.* 1995 Oct; 39(4): 407–410.

Uses: The pomegranate has a long history of herbal use dating back more than 3,000 years:

- Pomegranate extracts as potent antioxidants protect LDL cholesterol from oxidative damage, which allows it to damage the vascular walls.

- As a potent anti-inflammatory, pomegranate extract hinders inflammation that can lead to progression of atherosclerosis.

- The anti-platelet effect of the juice prevents intravascular clotting, which leads to ischemia.

- Experimentally, in animals, progression of plaques was prevented by pomegranate extracts.

- In a nonrandom Israeli study, 19 patients with carotid stenosis received a daily pomegranate drink. After a year, there was an improvement in the stenosis with a decrease in blood pressure. In the control group, stenosis was increased by 9%.

- In a randomized American study, 45 patients with coronary atherosclerosis were put on pomegranate drink 8 oz. daily for three months. The cardiac nuclear stress test showed improved blood flow compared with the control group.*

11. *Zygophyllum coccineum*

Synonyms: Zygophyllum, r'utrit, ratrayt, rotreyt, kammoon, karamaani, galam, gallam,

Active ingredients: Zygophyllin, quinovic acid, flavonoids

Medicinal parts: Fruits, seeds

---

* Pomegranates for the prostate and heart: Seeds of Hope. Harvard Men's Health Watch. 2007 Apr; 11(9): 4–5 (www.health.harvard.edu).

Actions: Anthelmintic. Alterative, diuretic, antipyretic, hemostatic

Uses: Stimulation of toad's heart[*] with zygophyllin and quinovic acid exhibited anti-inflammatory activity, cortisone-like action, choleretic and antipyretic activities. In rat studies from Kuwait, Gibbons and Oriowo found that the aqueous extract produced lowering in blood pressure, diuretic, antipyretic, local anesthetic and antihistaminic activities.[†]

Ebers Papyrus, as mentioned, has 842 prescriptions excluding the magical ones. The prescriptions utilize 328 different ingredients. In general, a prescription contained an average of 4.2 ingredients. Of all ingredients, 197 have not been translated (Estes). We have not begun to understand and utilize the plethora of herbology used by the ancient Egyptians. Further research in this field could be very beneficial.

ANCIENT EGYPTIAN MEDICINES USED BY AVICENNA

Malbus comminus

Coriandrum citrum (fresh and dried)

Allium sativum (garlic)

Lemon balm

Elbow amber

Indian agalloch

Lettuce seeds

Malabathrum

---

[*] Batanouny, K.H. and Ezzat, Nadia H. 1971. Eco-physiological studies on desert plants. I. Autecology of ZYGOPHYLLUM species growing in Egypt. *Oecologia* (Berl.), 7: 170–183.

[†] Gibbons, S. and Oriowo, M.A. Antihypertensive effect of an aqueous extract of Zygophyllum coccineum L. in rats. *Phytother Res.* 2001 Aug; 15(5): 452–455.

Indian spikenard

Imla Helenium

Green dates

Camphora officinarum

Santalum album

Quince (*Cydonia vulgaris*)

Borago officinalis

Doronicum scorpiodes

Musk

Amber

Cocoon of the silk

White Behmen

Cloves

Indian agalloch

Out of the 32 cardiological medications, I could translate 27 of them into English or Latin. Many of them are sedatives, tranquilizers, or hypnotics. Others are vasodilators or work against palpitation (see Kamal, 1975). It was very difficult to translate the remedies with the Egyptian pharmacopeia, ancient Egyptian plants, most of which have been Hellenized. However, it is an open field for research.

CHAPTER 15

Tribute to the Scribe

B EFORE CLOSING THIS MONOGRAPH, it is useful to quote from "The Immortality of Writers" (i.e. scribes), abstracted from P. Chester Beatty IV = P. British Museum 10684 (Estes and Kuhnke, 1984): (2, 5):

If you but do this you are versed in writings.

As to those learned scribes,
Of the time that came after the gods,
They who foretold the future,
Their names have become everlasting,
While they departed, having finished their lives,
And all their kin are forgotten.

They did not make for themselves tombs of copper,
With stelae of metal from heaven.
They knew not how to leave heirs,
Children [of theirs] to pronounce their names;
They made heirs for themselves of books,
Of Instructions they had composed.

They gave themselves [the scroll as lector]-priest,
The writing-board as loving son.
Instructions are their tombs,
The reed pen is their child,
The stone-surface their wife.
People great and small
Are given them as children,
For the scribe, he is their leader.

Their portals and mansions have crumbled,
Their *ka*-servants are [gone];
Their tombstones are covered with soil,
Their graves are forgotten.
Their name is pronounced over their books,
Which they made while they had being;
Good is the memory of their makers,
It is for ever and all time!

Be a scribe, take it to heart,
That is your name become (3, 1) as theirs.
Better is a book than a graven stela,
Than a solid ⌐tomb-enclosure.⌐
They act as chapels and tombs
In the heart of him who speaks their name;
Surely useful in the graveyard
Is a name in people's mouth!

Man decays, his corpse is dust,
All his kin have perished;
But a book makes him remembered
Through the mouth of its reciter.
Better is a book than a well-built house,

Than tomb-chapels in the west;
Better than a solid mansion
Than a stela in the temple!

If it was not for the scribes, from whom we inherited all of this knowledge, we would have been unaware of the magnificent Egyptian literature. We owe thanks to the scribes who preserved this knowledge from antiquity and saved it for eternity (see Figure 15.1).

FIGURE 15.1   Statue of Egyptian Scribe (c. Dynasty V) (commons.wiki-media.org).

Each tomb shapes in the very
Born than a solid garrison
Than whom the temple

that two, for the scribes, from which are prized all of
the knowledge, we would have been unaware of the magnit-
ent Egyptian Rite... We owe thanks to the scribes who preserved the knowledge from antiquity, and swell it he certainly
would have (f...)

FIGURE 6.1. Scribe at egg-tempera, the 21Dynasty. Y. Herrmann, with
red-ochre.

References

Abd-Ur-Rahman M.H. "The Four-Feathered Crown of Akhenaten", ASAE. 1959; 56:247–249 (Le Caire).

Al-Ghazal, S.K. "Ibn Al-Nafis and the Discovery of the Pulmonary Circulation", http://islamonline.net/English/Science/2002/08/article06.shtml#111:2004

Abdul Rahman, A *The History of Ancient Egyptian Medicine, Pharmacy and Chemistry* (Arabic). Cairo, 1939.

Allbutt, T.C. *Greek Medicine in Rome*. London: Macmilan, 1921.

Ammar, S. *En Souvenir de la Medecine Arabe*. Tunis: Basacone & Muscat, 1965.

Arberry, A.J. (trans.). *The Spiritual Physick of Rhazes*. London: John Murray, 1950.

Aristotle; Ross, W.D. (ed.). *Rhetorica*. Oxford University Press, 1959.

Arora, R.C. and Arora, S. "Comparative Effect of Cloflibrate, Garlic and Onion on Alimentary Hyperlipemia", *Atherosclerosis*. 1981(Jul); 39(4):447–452.

Bitschai, J. and Brodny, M.L. *A History of Urology in Egypt*. Riverside Press, 1956, 3–9.

Bonnabeau, R.C. "The School of Alexandria and the Vascular System", *Minnesota Medicine*. 1983(Feb); 66(3):100–101.

Borgstrom, G. *Fish as Food*. Volume 1. London: Academic Press, 1961.

Breasted, J.H. *A History of Egypt from the Earliest Times to the Persian Conquest*. London: Hadder and Stoughton, 1948.

Bridger, D. and Wolk, S. (eds). *The New Jewish Encyclopedia*. Springfield, NJ: Behrman House, Inc., 1995.

British Heart Foundation. "Coronary Heart Disease Statistics", 2002a.

British Heart Foundation. "European Cardiovascular Disease Statistics", 2002b.

Brovarski, E. et al. (eds.). *Egypt's Golden Age: The Art of Living in the New Kingdom, 1558–1085 B.C.* Boston: Museum of Fine Arts, 1982, 250–254.

Browning, R., Smith, M., and Jack, I. (eds.). *The Poetical Works of Robert Browning. Pauline and Paracelsus*, Vol. 1. London: Oxford University Press, 1983.

Brunner, H. et al. *Das Hertz im Umkreis des Glaubens*. Biberach, 1965.

Buckley, S.A. and Evershed, R.P. "Organic Chemistry of Embalming Agents in Pharaonic and Greco-Roman Mummies", *Nature*. 2001(Oct); 413:837–841.

Budge, W. *The Book of Medicines: Syrian Anatomy, Pathology, and Therapeutics*. London: Kegan Paul International Limited, 2000.

Cançado, J.R. Fian. Rev. Brasil di Biolog. 1944; 4:349.

Canton, R. "I-em-hotep, the Egyptian god of Medicine; Egyptian Views as to the Circulation", The Harveian Oration. *British Medical Journal*. 1904; 2269(5):1473–1476.

Cartwright, F. "The Mystery of Syphilis", in *Disease and History*. Marboro Books, 1991, 54–81.

Casson, L. and Kriegger, L. (eds.). *Ancient Egypt*. Alexandria, VA: Time Life Books, 1965.

Shorter AW Catalogue of Egyptian Religious Papyri in the British Museum (from XVIIIth to XXIInd Dynasty). London: The British Museum, 1938.

Cavallito, C.J. and Bailey, J.H. "Allicin, the Antibacterial Principle of *Allium sativum*. I. Isolation, Physical Properties and Antibacterial Action", *Journal of the American Chemical and Society* 1944; 66:1950–1954.

Champollion JF: *Précis du système héroglyphique des anciens Égyptiens*. Paris: Imprimerie royale, 1824.

Chapelain-Jaures, R. *La Pathologie dans l'Egypte Ancienne d'Apres les Momies* (these). Faculte de Medecine de Paris, 1920.

Chopra, D. *Quantum Healing*. New York: Bantam Books, 1990.

Clendening, L. *Lectures on the History and Philosophy of Medicine*. Lawrence: University of Kansas Press, 1952.

Clendening, L. *Source Book of Medical History*. Mineola, NY: Dover Publications, Incorporated, 1991.

Cockburn, A. et al. "Autopsy of Egyptian Mummy", *Science*. 1975(Mar); 187(4182):1155–1160.

Cooper, R. et al. "Trends and Disparities in Coronary Heart Disease, Stroke, and Other Cardiovascular Diseases in the United States Findings of the National Conference on Cardiovascular Disease Prevention", *Circulation*. 2000; 102:3137–3147.

Daremberg, C. et al. *Oeuvres de Rufus d'Ephese*. Paris: Imprimerie nationale, 1879.

David, A.R. (ed.). *Manchester Mummy Project*. Manchester University Press, 1979.

Dawson, W.R. *Science, Medicine and History*, (Charles Singer ed.). London: Oxford University Press, 1953.

Derry, H.E. and Engelbach, R. "Mummification", in *Annales du Service des Antiquités de l'Égypte*. 1942; volume 41, 233–265.

Diodorus Siculus. *Library of History, Vol. I: Books 1–2.34*. Translated by CH Oldfather. Loeb Classical Library 279. Cambridge, MA: Harvard University Press, 1933.

Doby, T. *Discoveries of Blood Circulation*. New York and London: Abel and Schuman, 1963.

Dragendorff, G. *Die Heilpflanzen der verschiedenen Völker und Zeiten*. Stuttgart: Ferdinand Enke, 1898.

Ebbell, B. (trans.). *The Papyrus Ebers: The Greatest Egyptian Medical Document*. Copenhagen: Levin and Munsgaard, 1937.

Ebers, G. and Stern, L. *Papyrus Ebers: Das Hermetische Buch Über die Arzeneimittel der Alten Aegypter*, 3 Vols. Leipzig: JC Henriches, 1875.

Edelstein, E.J. and Edelstein, L. *Aesclapius*. Baltimore, MD: Johns Hopkins University Press, 1945.

Edelstein, L. and Temkin, G.L. (trans.): *Ancient Medicine*. Baltimore, MD: Johns Hopkins University Press, 1967.

El Batrawi, A.M. *Mission Archeologique de Nubie 1929–1934: Report on the Human Remains*. Cairo: Government Press, 1935.

El-Gammal, S.Y. "The Role of Hippocrates in the Development and Progress of Medical Sciences", *Bull Indian Inst Hist Med Hyderabad*. 1993(Jul);23(2):125–136.

Eltorai, I.M. "History of Spinal Cord Medicine", in Lin, V.W., ed. *Spinal Cord Medicine*. New York: Demos, 2010: 999–1013.

Engelbach, R. *Introduction to Egyptian Archaeology*. Cairo: Imprimerie de l'Institut Francais d'Archeologie Orientale, 1946.

Erman, A. *Zaubersprache für Mutter und Kind*. Berlin Museum, Alshde: Kgl. Preuss. Akad. Wiss., 1901.

Erman, A.; Wild, H. (trans.). *La Religion des Egyptiens*. Paris: Perijot, 1937.

Estes, J.W. and Kuhnke, L. "French Observations of Disease and Drug Use in Late Eighteenth Century Cairo", *Journal of the History of Medicine and Allied Sciences*. 1984; (39):121–152.

Estes, J.W. *The Medical Skills of Ancient Egypt*. Canton, MA: Science History Publications, 1989.

Fayyad, S. (Arabic). *Ibn El-Nafis, the Discoverer of the Small Circulation (Pulmonary)*. Cairo: Al-Ahram Center for Scientific Translations, 1985.

Finlayson, J. "Ancient Egyptian Medicine", *British Medical Journal*. 1893; 748:1014, 1061.

Fleischauer, A.T. and Arab, L. "Garlic and Cancer: A Critical Review of the Epidemiologic Literature", *The Journal of Nutrition*. 2001; 131(Supplement):1032S–1040S.

Frazer, J.G. *Adonis and Osiris in the Golden Bough. A Study in Magic and Religion*. London: MacMillan and Co., 1936.

Ghalioungui, P. *Magic and Medical Science in Ancient Egypt*. London: Hodder and Stoughton, 1963.

Ghalioungui, P. *The Physicians of Pharaonic Egypt*. Cairo: Al-Ahram Center for Scientific Translations, 1983, 45.

Ghalioungui, P. "Four Landmarks of Egyptian Cardiology", *Journal of the Royal College of Physicians of London*. 1984(Jul); 18(3):182–86.

Ghalioungui, P. *The Ebers Papyrus: A New English Translation, Commentaries and Glossaries.*, Cairo: Academy of Scientific Research and Technology, 1987.

Goldberg, R.J. et al. "A Two-Decades (1975 to 1995) Long Experience in the Incidence, In-Hospital and Long-Term Case-Fatality Rates of Acute Myocardial Infarction: A Community-Wide Perspective", *Journal of the American College of Cardiology*. 1999; 33:1533–39.

Gordon, B.L. *Medieval and Renaissance Medicine*. London: Peter Owen, 1960.

Grapow, H. *Altagytischen Medizinichen Papyri*. Teil I. (Untersuchungen 1927, Mitteilung an der Vorderasiatisch – Aegyptisher Gessellschaft, Band 40, Heft I). Leipzig: JC Heinriches, 1935a.

Grapow, H. *Über die Anatomischer Kenntrisse den Alten Ägepteschen Ärzte*. Leipzig : JC Heinriches, 1935b.

Grapow, H. et al. *Grundriss der Medizin der Alten Aegypten*. Berlin: Akademe Verlag, 1955.

Gray, P.H.K. "Radiography of Ancient Egyptian Mummies", *Medical Radiography and Photography*. 1967; 43:34–44.

Grieve, M. *A Modern Herbal*. New York: Dover Publications, Inc., 1979.

Gruner, O.C. *A Treatise on the Cannon of Medicine of Avicenna*. London: Luzac and Co., 1930.

Harris, J.E. and Wente, E.F. (eds.). *An X-ray Atlas of the Royal Mummies*. University of Chicago Press, 1980.

Hart, G.D. et al. "Autopsy of an Egyptian Mummy (Nakht-Rom I)", *Canadian Medical Association Journal*. 1977; 117:461–473.

Hastings, J. *Encyclopedia of Religion and Ethics*, 6 Vols. Edinburgh: T & T Clark, 1925.

Heart and Stroke Foundation of Canada Website: www.heartandstroke.ca.

Helmuth, M.B. *Wonder Drugs, a History of Antibiotics*. Philadelphia, PA: Lippincott Williams & Wilkins, 2000.

Herodotus; Griffith, T. (ed.). *Histories*. Kent: Wadsworth Editions Limited, 1996.

Hirsch, A.T. et al. "Peripheral Arterial Disease Detection, Awareness, and Treatment in Primary Care", *JAMA*. 2001(Sep 19); 286(11):1317–1324.

Homer. *Odyssey*. New York: Barnes & Noble Books, 2003.

Hsing, A.W. et al. "Allium Vegetables and Risk of Prostate Cancer: A Population-Based Study", *Journal of the National Cancer Institute*. 2002(Nov 6); 94(21):1648–1651.

Hurry, J.B. *Imhotep, the Vizier and Physician of King Zoser and Afterwards the Egyptian God of Medicine*. Stoughton, MA: AMS Press, 1975.

Hussein, M.K. *The Edwin Smith Papyrus*. Cairo: Mondiale Press.

Hutchison. *Food and the Principles of Dietetics*. London: E. Arnold Ltd., 1956.

Iversen, E. Papyrus Carlsberg. No. VIII "in Det. K & L, Danske Videns Kabrines Selskab. Historisk; filologiske meddelser", XXVI, 5, Copenhagen, 1939.

Joachim, H. Papyrus Ebers, "Das Alteste Buch über Heilkunde". Berlin: JC Heinriches, 1890.

Jonckheere, F. *La Medecine Egyptienne*. Foundation Egyptologique Reine Elisabeth, Parc du cinquatenaire, 1944.

Jonckheere, F. *Le Papyrus Medical*. Brussels: Chester-Beatty, 1947.

Jonckheere, F. *Dans l'Arsenal Therapeutique des Anciens Égyptiens, Histoire de la Médecine 3*. Brussels: Chester-Beatty, 1953.

Kamal, H. *Ancient Egyptian Medicine* (Al-Tibb Almasri Al-Quadeem). Cairo: Al-Moqtataf and Al-Moquattum Publishers, 1922, 52.

Kamal, H. *A Dictionary of Pharaonic Medicine*. Cairo: National Publications House, 1967.

Kirchner, M. *Die Entdeckung des Blutkreislaufs*. Berlin Hirschwald, August, 1878.

Kreig, M. and Green, G. *Medicine, the Search for Plants that Heal*. London: George G. Harrap & Co., Ltd., 1965.

Kreitmair, H.E. *Merck's Jahresbericht*. 1956; 50:102.

Leak, H.E. "Ancient Egyptian Therapeutics", *Ciba Symposium*. 1940; 10:311.

Leake, C.D. *The Old Egyptian Medical Papyri*. Lawrence, KS: University of Kansas Press, 1952.

Lefebvre, G. *Essai sur la Medicine Egyptienne de l'Epoque Paraonique.* Paris: Presse Universitaire de France, 1956.

Lesho, E.P. et al. "Common Errors in Internal Medicine Peripheral Arterial Disease", *Federal Practitioner.* 2003(Jul); 20(7):52–60.

Lichtheim, M. *Ancient Egyptian Literature, A Book of Readings*, Vol. II: The New Kingdom. Berkeley, CA: University of California Press, 1976, 182–186.

Lloyd, J.U. *Origin and History of All Pharmacopeical Vegetable Drugs.* Cincinnati, OH: The Caxton Press, 1929.

Long, A.R. "Cardiovascular Renal Disease: Report of a Case Three Thousand Years Ago", *Archives of Pathology.* 1931; 12:92–94.

Magee, R. "Arterial Disease in Antiquity", *Medical Journal of Australia.* 1998(Dec); 169(11–12):663–666.

Major, R. "The School of Alexandria", in *History of Medicine*, Vol. 1. Springfield, IL: Charles C. Thomas, 1954; 141–151.

Manascha, I. "Die Gebursthilfe bei den Alten Aegypten", *Archives of Gynecology.* 1927; 131:425.

Manniche, L. *An Ancient Egyptian Herbal.* British Museum Press (Distribution); 1st University of Texas Press edition, December 31, 2006.

Menon, S.I. et al. "Effect of Onions on Blood Fibrinolytic Activity", *British Medical Journal.* 1968; 3:351–352.

Monro, T.K. *The Physician, A Man of Letters, Science, and Action.* Edinburgh: Livingstone, Ltd., 1951.

Moon, R.O. *The Relation of Medicine to Philosophy.* London: Longman, Green & Co., 1909.

Moursi, H. *Die Heilpflanzen im Land der Pharaonen: ägyptisch-nubische Volksmedizin.* Kairo: Lehnert & Landrock, 1992.

Mutter, H. "Behandlung von Krebskranken in landarzlicher Parxis", *Erfahrungsheilkunde.* 1965 Heft 10 und 1968 Heft 1.

National Center for Health Statistics: http://www.cdc.gov/nchs/about/major/dvs/mortdata.htm, 2004.

Nazmi, A.A. *La Medecine au Temps des Pharaons.* (These) Montpelier: Faculte de Medecine, 1903.

Neuburger, M. *History of Medicine.* London: Oxford University Press, 1910.

Nunn, J.F. *Ancient Egyptian Medicine.* University of Oklahoma Press, 1996 and 2002.

Oakes, L. and Gahlin L. *Ancient Egypt: An Illustrated Reference to the Myths, Religions, Pyramids and the Temples of the Land of the Pharaohs.* Hermes House, 2002.

Olms, G. *Avicenna, Liber Canonis.* Hildesheim, 1966.

Osler, W (with quote from Fuller). "A Note on the Teaching of the History of Medicine", *British Medical Journal*. 1902; 2: 93.

Osler, W. *The Evolution of Modern Science*. New Haven: Yale University Press, 1923.

Paré, A Les Oeuvres. (29 volumes). Paris: G. Buon, 1598; 528, 554, 559.

Piankoff, A. *Le Coeur dans les Textes Egyptiens*. (These presentee pour le Doctorat de l'Universite de Paris). Paris: Librairie Orientaliste, Paul Genthmer, 1930.

Posener G: *Dictionary of Egyptian Civilization*. New York: Tudor Publishing Co., 1959.

Reisner G.A. *The Hearst Papyrus*. Hieratic text. Leipzig: JC Heinriches, 1905.

Rodwell, J.M. (trans.). *The Koran*. New York: Bantam Books, Incorporated, 2004.

Roland, L. "Vasculitis Syndromes", in Rowland L, ed. *Merritt's Neurology*, 10th ed. Philadelphia, PA: Lippincott Williams & Wilkins, 2000: 905–906.

Roth, C. *The Jewish Contribution to Civilization*. London: MacMillan & Co., 1938.

Ruffer, M.A. and Moodie, R.L. (eds.). *Studies in Paleopathology of Egypt*. The University of Chicago Press, 1921.

Rühli, F.J. et al. "Clinical Magnetic Resonance Imaging of Ancient Dry Human Mummies Without Rehydration", *JAMA*. 2007; 298(22): 2618–2620.

Ryle, A.J. *The Natural History of Disease*. London: Oxford University Press, 1948.

Sainani, G.S. et al. "Onion, Garlic, and Experimental Atherosclerosis", *Japan Heart Journal*. 1979(May); 20(3):351–57.

Sallmann, J. *A Manual of Pharmacology*. Philadelphia, PA: WB Saunders Co., 1957.

Sandison, A.T. "Degenerative Vascular Disease in the Egyptian Mummy", *Medical History*. 1962; 6:77–81.

Sandison, A.T. "Diseases in Ancient Egypt", in Cockburn A, Cockburn E, eds. *Mummies, Disease, and Ancient Cultures*. Cambridge University Press, 1980.

Sarton, G. *Introduction to the History of Science*. Krieger Publishing Company, 1975.

Schering Corporation. *Medicine and Pharmacy: An Informal History. 1: Ancient Egypt. Schering Corporation*, 1975.

Seiner, R. *Egyptian Myths and Mysteries*. London: Percy Lund – Hymphries & Co., 1933.

Shattock, S.G. "Microscopic Sections of the Aorta of King Mernaphtah", *Lancet*. 1909(Jan); 1:319.

Shaw, A.F.B. "A Histological Study of the Mummy of Har-mosĕ, the Singer of the Eighteenth Dynasty (c. 1490 BC)", *Pathology and Bacteriology*. 1938; 47:115–123.

Silverman, D.P. (General Editor). *Ancient Egypt*. New York: Oxford University Press, 1997.

Singer, C. *Studies in the History and Method of Science*. Oxford: Clarendon Press, 1917.

Singer, C. and Underwood, E.A. *A Short History of Medicine*. Oxford: Clarendon Press, 1962.

Smith, E. "The Unwrapping of Pharaoh", *BMJ*. 1908; 1:342–343.

Stauffer, R.C. *Science and Civilization*. Madison, WI: University of Wisconsin, 1943.

Stetter, C. *The Secret Medicine of the Pharaohs: Ancient Egyptian Healing*. Chicago, IL: Edition Q, 1993.

Stever, R.C. and Saunders, J.B. *Ancient Egyptian and Indian Medicine*. Berkley, CA: University of California Press, 1959.

Stoll, A. and Seebeck, E. "The Specificity of the Alliinase from Allium Sativum", *Comptes rendus de l'Académie des Sciences*. 1951(Apr 9); 232(15): 1441–1442.

Thacker, P.H. et al. "Chronic Stress Promotes Tumor Growth and Angiogenesis in a Mouse Model of Ovarian Carcinoma", *Nature Medicine*. (2006); 12:939–944.

Trease, G.C. *A Textbook of Pharmacognosy*. Lord Balliere, Tindall & Cassel, 1966.

Twentyman, L.R. "The Mistletoe and Cancer", *The British Homeopathic Journal*. 1954(Apr).

Van Praagh, R. and Van Praagh, S. "Aristotle's 'Triventricular' Heart and the Relevant Early History of the Cardiovascular System", *Chest*. 1983(Oct); 84(4):462–468.

Vatsala, T.M. et al. "Effects of Onion in Induced Atherosclerosis in Rabbits: I. Reduction of Arterial Lesions and Lipid Levels", *Artery*. 1980; 7(6):519–530.

Wallis, T.E. *A Textbook of Pharmacognosy*. London: JA Churchill Ltd., 1967.

Watt, J.M. and Gerdina, M. *The Medical and Poisonous Plants of Southern and Eastern Africa*. Edinburgh: Breyer-Brandivijk, LS Livingsone Ltd., 1962.

Weigall, A. *The Life and Times of Akhnaton, Pharaoh of Egypt*. London: T. Butterworth, 1922.

Willerson, J.T. and Teaff, R. "Egyptian Contributions to Cardiovascular Medicine", *Texas Heart Institute Journal*. 1996; 23(3):191–200.

World Health Organization (WHO) CVD Strategy. "Cardiovascular Diseases—Prevention and Control", 2001/2002.

World Health Organization (WHO). "Prevention of Cardiovascular Disease: Guidelines for assessment and management of cardiovascular risk", World Health Organization, Geneva, 2007.

World Health Organization Website (WHO): www.who.int/ncd/cvd, 2004.

World Health Organization Website (WHO) about cardiovascular disease: https://www.who.int/cardiovascular_diseases/en/.

Wren, R.C. *Potter's New Cyclopaedia of Botanical Drugs and Preparations*. Woodstock: Beekman Publishers Inc., 1994.

Wreszinski, V.W. *Der Grosse Medizinische Papyrus*. Des Berliner Museums. Papyrus Berlin 3038. Leipzig: JC Heinrichhe, Buchhandlung, 1909.

Wreszinski, V.W. *Der Londoner Medizinische Payprus*. British Museum No. 10059 und Der Papyrus Hearst. Leipzig: JC Hinrich'she, Buchhandlung, 1912.

Wreszinski, V.W. (ed.). *Der Papyrus Ebers: Umschrift, Übersetzung und Kommentar*. Leipzig: JC Heinriches, 1913.

Yamagita, T. et al. "Cycloalliin, a Cyclic Sulfur Imino Acid, Reduces Serum Tricylclycerol in Rats", *Nutrition*. 2003(Feb); 19(2):140–143.

Zhou, B. et al. "Overweight is an Independent Risk Factor for Cardiovascular Disease in Chinese Populations", *Obesity Review*. 2002 (Aug); 3(3):147–156.

Zimmerman, M.R. "The Mummies of the Tomb of Nebwenenef: Paleopathology and Archeology", *Journal of the American Research Center in Egypt*. 1977; 14:33–36.

Zimmerman, M.R. "The Paleopathology of the Cardiovascular System", *Texas Heart Institute Journal*. 1993; 20:252–257.

Zink, A.R. et al. "Molecular Study on Human Tuberculosis in Three Geographically Distinct and Time Delineated Populations from Ancient Egypt", *Epidemiology and Infection*. 2003; 130(2): 239–49.

Ziskind, B. and Halioua, B. "La conception du cœur dans l'Egypte ancienne", *Medicine/Sciences*. 2004(Apr); 20:367–373. (French).

Appendix: Book Reviews

BOOK REVIEW OF THE PREVIOUS EDITION BY SIR ZACHARY COPE

This is a typed manuscript about 140 pages dealing with the medical history of ancient Egypt. The author is a distinguished surgeon and a Professor of Surgery in the University of Cairo. He speaks Arabic, English, French, German, and Italian and appears to have a good knowledge of the literature in those languages.

In the Introduction, he remarks on the variations in the causes of death in different periods of history and expresses a wish to find out whether cardiovascular disease was a common disease in ancient Egypt, especially in Greco-Roman times.

After a very brief reference to the history of Egypt there is inserted an excellent chronological list of the ancient historical periods according to the latest authorized dates. He then tells of the early influence of magic and religion upon medical practice and traces the gradual development of a knowledge of anatomy and of physiology; knowledge of anatomy was helped by the practice of embalming, which is briefly described, while physiology must have begun by the observation that a pulse-wave was felt all over the body after a heartbeat.

All the known ancient Egyptian medical papyri are mentioned and some cases commented upon, especially the famous Ebers Papyrus.

Two very interesting sections deal respectively with the pathology and pharmacology of those times. Some ancient specimens show that tuberculosis of the spine and that degeneration and calcification of the larger arteries were common, at least during the Greco-Roman periods. Cardiovascular disease was therefore probably as common then as it is today.

The drugs in use in ancient Egypt were numerous and many are still in use today. A large list of them is given. The author thinks it likely that further research might reveal important active principles in some of them. Many spices prevent decomposition and must have contained antiseptic properties.

This little treatise might serve well as "A Short Introduction to Ancient Egyptian Medicine". It could be made more attractive by the addition of a few more illustrations, and a bibliography should be added and a full index provided.

Sir Zachary Cope
Reviewed, 1956

BOOK REVIEW OF THE PREVIOUS EDITION BY KHALID EL-DISSOUKY, PH.D.

Many books dealing with ancient Egyptian medical science and knowledge have been published in recent years, and many Egyptologists have written much more, submitting the hieratic texts to detailed analyses where philology and medicine meet, and they have thereby presented valuable syntheses of the subject; but none of all the above-mentioned works treating the subject of angiology and the vascular system as known in ancient Egyptian medicine meticulously and scientifically as done in the present work. As a surgeon whose work is mostly done in connection with cardiovascular diseases, Dr. Ibrahim Eltorai succeeded in this book in illustrating the fact that the ancient Egyptians understood the physical as well as the moral functions of the heart. Moreover, after a thorough study of the ancient *materia medica*, particularly those concerning heart and vascular disease, he proved that the ancient Egyptians had a considerable knowledge of angiology. An instructive comparison has been made between the ancient and the new notions, which sheds new light on how far our ancestors reached in this field.

Dr. Eltorai's comments on the various texts are noteworthy achievements in themselves. He has carefully treated the sections through informative discussions, exploring various avenues of approach and suggesting alternative explanations where there are problems of interpretation. In fact, this is a pioneering work in reading and giving new interpretations to the sundry texts dealing with the vascular system, as well as the cardiovascular diseases. This work achieved by Dr. Eltorai invites us to be grateful for this new contribution to the universal history of sciences.

Khalid El-Dissouky, Ph.D.
Ph.D. Chicago
Professor of Egyptology
Ain Shams University, Cairo
Reviewed, 1954

BOOK REVIEW OF THIS EDITION BY SIR MAGDI YACOUB, F.R.C.S., L.R.C.P., F.R.S.

A copy of the book entitled *A Brief History of Ancient Egyptian Medicine: with Emphasis on Cardiovascular Diseases* was sent to me by Ibrahim Eltorai, M.D., a colleague of mine at the Faculty of Medicine and formerly a vascular surgeon at the same school at Cairo University. He requested that I review this second edition. After a fast review, I found that the book contains 16 chapters. The chapters, as in the title, cover ancient Egypt's history, its cultures, its religions, its philosophy, embalmment, and its dynasties. It has a brief outline of ancient Egyptian medicine, particularly anatomy, physiology, pathology, clinical practice, especially ancient therapeutics. All medical papyri were presented, but the author focused on the world's oldest book of medicine, i.e. the Ebers Papyrus. Because of his medical background, Dr. Eltorai elaborated his studies on cardiovascular medicine from all its aspects. He came out with interesting findings in this field and put them in a nutshell, making this knowledge easy to obtain.

I found out that the text was reviewed by two Egyptologists and the first edition by the late Sir Zachary Cope. I will not hesitate to recommend it for medical students to learn the splendid knowledge of their ancestors and to look into botanical management of cardiovascular diseases that might be a field of investigation.

Sir Magdi Yacoub, F.R.C.S., L.R.C.P., F.R.S.
Reviewed, 2014

BOOK REVIEW OF THIS EDITION
BY FAYZA HAIKAL, D.PHIL.

It is my pleasure indeed to review Dr. Eltorai's historical account of ancient Egypt for medical students as a background for his book on ancient Egyptian medicine since, as he says himself in the introduction of the book, "it is impossible to follow the history of science without knowing the chronology of those old times". Dr. Eltorai summarized ancient Egypt's long history in a few paragraphs giving the reader just the essentials that will help put Egypt in its historical context and allow him to appreciate the development of this history over 5,000 years, its international contacts through its expansion in the Middle East and in the heart of Africa, with its moments of glory and great creations, but also through the vicissitudes of less happy times during its long and rich history. But regardless of the political situation, Egypt remained always famous in the whole region for its medical skills, and this vast medical knowledge recorded on papyrus can be sensed even in its religious hymns to the creator god.

Dr. Eltorai reminds us in his introduction to the book that, as man remains the same, the fundamental problems of disease are often the same. Therefore early ideas in medicine should be revisited every now and then to see if the knowledge of the past cannot be reformulated into modern scientific principles. And since physicians' aim is to treat their patients' ailments, these should be given "a remedy, effective as much as possible, no matter it is made in the East or West, North or South, for it is all man's world".

Dr. Eltorai's brief history of ancient Egypt gives the reader a feeling of the time. It also underlines the recurrence of historical events and the interaction of the past with the present. The list of important museums in the world and the selected bibliography may also give the reader a better appreciation of Egypt's legacy and diffusion in the world.

Fayza Haikal, D.Phil.
Professor of Egyptology at
The American University in Cairo.
Reviewed, 2012

Printed in the United States
by Baker & Taylor Publisher Services